Combatting Climate Change in the Pacific

Marc Williams • Duncan McDuie-Ra

Combatting Climate Change in the Pacific

The Role of Regional Organizations

Marc Williams
School of Social Sciences
UNSW Sydney
Sydney, NSW, Australia

Duncan McDuie-Ra
School of Social Sciences
UNSW Sydney
Sydney, NSW, Australia

ISBN 978-3-319-88816-3 ISBN 978-3-319-69647-8 (eBook)
https://doi.org/10.1007/978-3-319-69647-8

Softcover re-print of the Hardcover 1st edition 2018

Printed on acid-free paper

This Palgrave Macmillan imprint is published by Springer Nature
The registered company is Springer International Publishing AG
The registered company address is: Gewerbestrasse 11, 6330 Cham, Switzerland

Acknowledgements

We would also like to thank the research assistants who worked on this project (Rebecca Pearse, Henar Perales, and Charlotte Penel); the many representatives from regional and international organizations, NGOs, and governments in the Pacific who gave us time and valuable insights, in particular David Sheppard (SPREP); and those colleagues who provided feedback at the various seminars and conferences where we have presented this work, particularly Carola Betzold whose comments were especially useful.

Contents

1 Introduction: The Politics of Climate Change in the Pacific 1

2 Organizing a Regional Response to Climate Change in the Pacific 13

3 Constructing Climate Change in the Pacific 39

4 Constructing Climate Security in the Pacific 63

5 Organizing Climate Finance in the Pacific 87

6 Conclusion: The Future of Climate Politics in the Pacific 109

References 117

Index 133

List of Acronyms

ADB	Asian Development Bank
AOSIS	Alliance of Small Island States
CDM	Clean Development Mechanism
COP	Conference of Parties of the UNFCCC
CROP	Council of Regional Organisations in the Pacific
FAO	Food and Agriculture Organization
FFA	Pacific Islands Forum Fisheries Agency
FRDP	Framework for Resilient Development in the Pacific
GCF	Green Climate Fund
GEF	Global Environment Facility
IUCN	International Union for the Conservation of Nature
MIRAB	Migration, remittances, aid dependency and bureaucracy
NAPA	National Adaptation Programme of Action
OECD	Organisation for Economic Co-operation and Development
PICTs	Pacific Island Countries and Territories
PIDP	Pacific Islands Development Program
PIF	Pacific Islands Forum
PIFACC	*Pacific Islands Framework for Action on Climate Change*
PIFS	Pacific Islands Forum Secretariat
PIGGAREP	Pacific Islands Greenhouse Gas Abatement through Renewable Energy Project
SIDS	Small Island Developing States
SOPAC	Pacific Islands Applied Geoscience Commission
SPC	The Pacific Community, formerly the South Pacific Commission
SPREP	Secretariat of the Pacific Regional Environment Programme

UNDP	United Nations Development Programme
UNEP	United Nations Environment Programme
UNESCAP	United Nations Economic and Social Commission for Asia and the Pacific
UNFCCC	United Nations Framework Convention on Climate Change

List of Figures

Fig. 5.1	Financial flows for climate change mitigation and adaptation in developing countries	89
Fig. 5.2	Global climate finance pledges according to the type of source	90
Fig. 5.3	Approved and disbursed climate finance in Pacific by type of administering agency	97
Fig. 5.4	Climate finance by stated purpose of fund	99
Fig. 5.5	Climate finance by focus of national pledges (adaptation, mitigation general, mitigation REDD)	100
Fig. 5.6	Climate finance by approved individual projects	101

List of Tables

Table 2.1 Secretaries General of the PIFS 22
Table 2.2 Directors General of the Pacific Community 27
Table 2.3 Executive Heads of SPREP 30
Table 5.1 Sources of climate finance (funds, administrating bodies, organizational type, and stated purpose) 91
Table 5.2 Multilateral Funds Supporting Adaptation, 2003–2013 (US$ millions) 93

CHAPTER 1

Introduction: The Politics of Climate Change in the Pacific

Abstract The introductory chapter makes the case that within the grand architecture of global climate governance, analysing regional complexes in the Pacific provides new insights into the ways climate change is constructed, governed, and shaped by—and in turn shapes—regional and global climate politics. Three claims are made. First, the Pacific is not just 'any region', rather the Pacific has been constructed as the frontline of climate change. Second, climate change reinforces the notion of regional solidarity in the Pacific institutionalized in regional organizations; however, these organizations have become heavily dependent on external donors in combatting climate change. Third, Pacific states have advocated for important changes to the global architecture of climate finance, yet contestation over key elements of climate finance leaves the region poorly served.

Keywords Climate change • Pacific regionalism • Climate governance • Aid dependency

From the perspective of Pacific Island countries and territories (PICTs) much of the intense speculation and debate over climate change action, policy, and governance seems irrelevant and immature given the oft-repeated warnings of the severe consequences of climate change for low-lying islands and atoll countries. The Pacific is not just 'any region' when

M. Williams, D. McDuie-Ra, *Combatting Climate Change in the Pacific*, https://doi.org/10.1007/978-3-319-69647-8_1

it comes to climate change but rather, as Farbotko (2010) argues, the Pacific is an 'experimental space' of climate change 'canaries'. Representatives from PICTs speak about the coming catastrophe and the impacts already being felt at virtually every governmental and non-governmental meeting on climate change. Small states from the region have been able to intervene in international negotiations, to exert emotional pressure, to keep the symbolism of societies under threat of extinction at the forefront of global consciousness on climate change. Advocates and activists from all over the world frequently invoke the region when making a case for climate action. Images of Pacific Islanders knee deep in water adorn websites, posters, and campaign materials from university campuses to large international non-governmental organization headquarters. Experts and consultants have descended on the region to address adaptation, mitigation, and policy adjustments in PICTs. The position of the Pacific at the frontline of climate change has spurred voluminous research; however, studies of the politics of climate change within the region is limited, particularly with regard to regional processes and the ways climate science, climate security, and climate action are internalized, debated, and acted upon in the region itself.

While it is widely agreed that climate change requires action at multiple levels of governance, studies of climate change in the Pacific have been narrowly focused with limited attention to transnational and regional processes. There has also been limited attention to how the shared fate of climate change has created new solidarities, a new sense of the region, new vulnerabilities and dependencies, and new forms of agency. Yet at the same time climate change has come to dominate the region's politics, aid and finance, and development agenda. Exploring the regional politics of climate change in the Pacific draws our attention not only to political dynamics in the region, but within the grand architecture of global climate governance the region provides, compelling new insights into the ways climate change is constructed, governed, and shaped by—and in turn shapes—regional and global climate politics. Our focus has two strands. The first is to explore the ways climate change is constructed in the Pacific, first as an environmental problem and second as a security threat (Chaps. 3 and 4). While we are interested in localized versions of this construction within PICTs—detailed in Chap. 3, it is the regional level that holds most resonance for our understanding as this aligns with the ways climate governance is organized and financed. The second strand is to sketch the structure of regional governance, regional finance, and the dynamics shaping these through the last decade of climate politics. By focussing on these two strands, namely the ways climate change and 'climate security' are constructed in the Pacific and the ways the concept mobilizes resources and shapes the implementation of climate finance, this book provides an account of the way regional organizations in the Pacific have contributed to the search for solutions to the problem of climate change. Our analysis brings to the fore competing conceptions of climate security, the articulation of policy narratives, and the constraints imposed by continued dependence on external powers. Through an exploration of regional governance as a strategy by which small vulnerable states respond to urgent crises, the book explores both the potential and the limitations of collective action on environmental issues.

Throughout the book we make three main arguments. First, the Pacific is constructed as the frontline of climate change. These narratives are internalized and have come to dominate policy making at the regional level, in specific PICTs, and in the ways powerful regional actors, both multilateral and bilateral, approach climate change. We identify two duelling discourses: insecurity and vulnerability. These coexist and PICTs draw from both when necessary in their attempts to shape the ways climate change is addressed in the region and when engaging different audiences.

Second, climate change reinforces the notion of regional solidarity in the Pacific institutionalized in three key regional organizations. Given the limited capacity of PICTs in terms of knowledge gaps, institutional deficiencies, and limited financial resources, a regional response appears a logical way of addressing the impacts of climate change. These organizations mobilize resources for climate change action through distinct policy narratives about the place of the Pacific in the global climate crisis. However, beneath the surface regional climate governance is heavily dependent on donors for finance and expertise. PICTs are thus extremely vulnerable to the impacts of climate change and extremely vulnerable to growing dependencies on donors to address these impacts.

Third, PICTs have not simply been passive recipients of climate finance but have exercised their agency to demand changes to the existing multilateral climate finance architecture. Despite this, the failure to reach agreement over the objectives of climate finance, the required volume of finance, and the governance of climate finance negatively impacts the ability of small island states in the Pacific to secure adequate finance in order to combat climate change at various scales in the region and domestically.

Pacific Island States and Territories

Our research focused on the Pacific, which can be a geographically vague term. The challenge with the term is determining which states and territories are included and which are excluded. There is little consensus as can be seen in membership of regional organizations in the Pacific. For instance, the Pacific Islands Forum (PIF), the flagship intergovernmental organization in the region, restricts membership to independent states, and has 18 members (including Australia and New Zealand), though it also includes several 'associate members'. The [1]Pacific Community (SPC)—the oldest regional organization initially started by colonial powers prior to independence movements in the region—has 26 members and includes many non-independent territories, such as French Polynesia, Tokelau, and Guam. It also includes founding regional and extra-regional powers such as the United States and France (the Netherlands and the United Kingdom are no longer members). Thus, while the SPC is more inclusive to non-independent territories it is also inclusive of colonial and neo-colonial powers. Colonial relationships have a major role in inclusion and exclusion; even still there are some seemingly curious cases of inclusion and exclusion, especially when it comes to internal borders. Papua New Guinea is a member of most Pacific regional organizations, while Indonesia—despite sharing a land border with the same landmass—is not. Timor-Leste, despite sharing many of the same vulnerabilities and having a close—though not always easy—relationship with Australia, has instead been pushing for membership of the Association of South East Asian Nations (ASEAN), though it is a member of the Alliance of Small Island States (AOSIS). AOSIS is a global grouping of small island states, and has 14 members from the Pacific; all independent states except the Cook Islands and Niue. Australia, New Zealand, the United States, and France are not members.

Inclusion and exclusion based on sovereign status make it challenging to find a catch-all term that we can use throughout the book for independent and non-independent states. A related problem is that the idea of 'the Pacific' or 'the South Pacific' is a creation of what van Schendel refers to as 'the scramble for area' in post-Second World War knowledge production in universities and among governments and multilateral institutions (2002: 647–8). This knowledge production naturalizes the idea that Southeast Asia ends at Indonesia's border with Papua New Guinea or that Palau is part of the Pacific while the Philippines is not. A pertinent question for us during our research was whether there is anything connecting territories as disparate as the Marshall Islands and New Caledonia. Or Papua New Guinea and Tonga. Is a common grouping forced? Is it just habitual given the existence of institutions that group these states and territories together? Or of the existence of area-specific knowledge that designates this as a region? While we are conscious of this problem, we found that climate change, climate security, and the ways these have been addressed has built, or strengthened, a sense of commonality shared by these states and territories. We are comfortable that a notion of 'the Pacific' as a region exists and is accepted by polities and communities subject to the term, to the idea.

Furthermore, given that several powerful 'developed' states are included in regional groups, especially Australia and New Zealand, what are we referring to when we talk about the Pacific? We have used PICTs, an acronym common in the region itself (indeed this is when we first encountered it), throughout the book. This solves one problem by accommodating independent states and overseas territories under varied governing arrangements. But it does not solve the second problem of whether we include Australia and New Zealand (or even France and the United States), though we include their overseas territories. This gets complicated when we talk about the PICTs as a collective. For instance, we may refer to the donor dependency in PICTs, yet the donor in this case may be Australia, which is a member of the three key institutions in the region, institutions that funnel donor funds to other members. To simplify, for the most part we use PICTs to refer to independent states and territories in the Pacific excluding Australia, New Zealand, France, and the United States, except when we mention these states specifically or refer to positions by organizations of which they are members. That leaves us with the following PICTs: American Samoa, Cook Islands, Federated States of Micronesia, Fiji, French Polynesia, Guam, Kiribati, Marshall Islands, Nauru, New Caledonia, Niue, Northern Mariana Islands, Palau, Papua New Guinea, Pitcairn Islands, Samoa, Solomon Islands, Tokelau, Tonga, Tuvalu, Vanuatu, and Wallis and Futuna. We also believe that this is in concert with views in the Pacific we encountered during fieldwork, which tended to exclude Australia, New Zealand, France, and the United States from collective identities, even if they were included in institutions. We recognize this group, this region, is constructed—a region created from various colonial, neo-colonial, and decolonization processes—but despite fissures between members and subgroupings, this grouping of polities and communities shares a sense of solidarity, of shared fate, of a way of living in a very loose sense, that makes it a region. We argue throughout the book that the challenge of climate change bolsters a sense of regionalism among PICTs, and relationships of solidarity outside the region through AOSIS. To put it another way, climate change, climate insecurity, and the action taken to combat the impacts is shaping the region in profound ways, despite the historical fragility of its construction. Indeed, the shared ecology of PICTs experiencing climate change appears to be ushering in new notions of borders in the Pacific.

Approach and Methods

Combatting Climate Change brings together two scholars from different disciplines: International Relations (Williams) and Development Studies (McDuie-Ra). Utilizing our different backgrounds, we have adopted a three-part approach to the study.

First, we seek a comprehensive understanding of the role of regional organizations in developing climate governance in the Pacific. To gain the necessary in-depth knowledge we undertook fieldwork in the Pacific on numerous trips between 2009 and 2012 to Fiji, Papua New Guinea, New Zealand, Samoa, the Solomon Islands, and Vanuatu in addition to work in Australia where we are both based. During these trips we undertook face-to-face interviews with key individuals in regional organizations including the Secretariat of the Pacific Regional Environmental Program (SPREP), the Pacific Islands Applied Geoscience Commission (SOPAC), and with individuals working on climate change in international and regional organizations operating in the Pacific including the PIF, the United Nations Development Program (UNDP), and United Nations Economic and Social Commission for Asia and the Pacific (UNESCAP). We visited some of these organizations more than once to speak to staff in different portfolios and departments. We conducted interviews with civil society organizations advocating for climate change action and other environmental issues in the region. We also met with researchers and consultants based in the region, in particular at the University of the South Pacific in Fiji. We conducted the majority of this work together, with the exception of trips to Papua New Guinea (Williams) and the Solomon Islands (McDuie-Ra).

Interviewees generally navigated a balance between official transcripts and some candid insights. As all interviewees agreed to interviews on conditions of anonymity we have quoted few interviewees at length.

Furthermore, the aim of our interviews was not to 'catch out' representatives of these organizations saying things that were otherwise unavailable publicly, but rather to understand the mechanics of regional organization and locate additional material to add further depth. In addition to interviews, fieldwork granted access to documentation and grey literature unavailable outside the region. This material was crucial to the arguments developed in this book and we were fortunate to have a number of research assistants working through this vast corpus of material through 2013–15.

Second, we adopt a critical methodology underpinned by a constructivist approach. In exploring the nexus between knowledge and power the book foregrounds the reasons behind the emergence of climate change as the dominant issue in current understandings of environmental insecurity in the Pacific. To understand how climate change in the Pacific is constructed, how these constructions are internalized and translated, how they are challenged and critiqued, and how they shape and are shaped by policy implemented by national governments and regional organizations, we engaged in a critical analysis of various policy narratives developed by scholars, scientists, consultants, governments, and bureaucrats. The data corpus of this analysis is based on a review of the extant literature, documents from governments—with a particular focus on national climate actions plans—regional organizations, and our interviews. A critical reading of these texts reveals the different ways in which climate change in the Pacific has been conceptualized and the ways different narratives have influenced policy ideas and implementation. In this context, we define a policy narrative as a narrative or story about a specific policy problem designed to influence decision-making, and shape policy implementation. These narratives create shared meanings for decision-makers both within and outside of a polity who are concerned with the same problem set. Thus in the case of climate change in the Pacific these policy narratives link decisions at the local, national, regional, and transnational levels. When a set of narratives takes a recognizable shape we refer to it as a discourse.

Third, as the book provides the first detailed examination of the role of regional environmental cooperation in the Pacific, we undertook the unglamorous task of describing the regional political and financial architecture. This was a surprisingly complex task. Multiple organizations, donors, overlaps, and changing political circumstances during our period of research made this much more difficult than we had first imagined. This also gave us a sense of how challenging regional environmental cooperation is for governments in the Pacific. The number of schemes, donors, and requirements are difficult to establish and monitor. This is compounded by the long periods of travel to attend international meetings, the lack of human resources in many government ministries and regional organization departments, the impacts of climate-related extreme weather events on infrastructure and mobility, and the career trajectories of many professionals in climate and environmental roles. Frequently during our research we would return to the region to visit a contact only to find they had gone to another organization, into government, or into consulting. We even conducted an interview with the same individual in two different organizations after she changed roles between our visits. The annual United Nations Framework Convention on Climate Change (UNFCCC) Conference of the Parties (COP) meetings mean almost perpetual preparation for the next meeting among members of regional organizations and governments, in addition to regional meetings and meeting with donors. Therefore we have attempted to assemble a clear understanding of the regional architecture as it developed during the period of research for this book, recognizing the prospects for change are constant.

Structure of the Book

Throughout the book we draw upon two strands mentioned above—the ways climate change is constructed as an environmental problem and as a security threat, and the structure of regional governance and regional finance that contends with these challenges—to explore the interplay between discourses of climate change and political action taken to combat it. The following chapter, 'Organizing a Regional Response to Climate Change in the Pacific' explores the origins of contemporary regionalism in the Pacific. We focus on the degree of regional interdependence, the strength of regional identity or regional awareness, and the creation of regional organizations that reflect and in some cases create this identity. We argue that climate change has shaped regional organization, in part out of necessity and in part by calling upon existing regional solidarity and the notion of shared fate. The first part of the chapter analyses the sense of shared identity drawn from the colonial experience, the nature of decolonization (and its resistance) in the region, and the articulation of a shared regional identity—a regional culture that underpins regional politics. The 'Pacific Way', an imprecise concept with a broad uptake, generally refers to a culturally derived approach to conflict resolution and a negotiating tool based on consensus and compromise. The Pacific Way is the ideational underpinning of key regional organizations and a way of structuring decision-making in a way that reinforces distinctiveness and a shared way of approaching cooperation and contention. We explore the birth of the concept and the ways it is interwoven into the operations of the three key regional organizations that we will refer to throughout the book: the PIF, the SPC, and the Secretariat of the Pacific Regional Environment Programme (SPREP). The second part of the chapter maps the mandates, resources, and competencies of the various regional organizations concerned with climate change in the Pacific, illustrating both the complexity and coherence in the regional architecture.

The prominence of climate change in the global arena has reconfigured the Pacific and its place in the world, the focus of Chap. 3, 'Constructing Climate Change in the Pacific'. Climate change has constructed the Pacific as the frontline of climate change. The first part of this chapter traces the emergence of climate change narratives and how they have been contextualized in the region and have established hegemony in political arenas. We focus on six narratives of climate change in the Pacific: sea level rise, possible extinction, livelihoods, gender, and security—the latter is taken up in more detail in the following chapter. These narratives create shared meanings for decision-makers and link decisions on combatting climate change at the local, national, regional, and transnational levels. We follow this with an analysis of critical voices—many from within the region—that challenge the hegemony of climate change in local politics and the implications for other environmental and development problems faced in the region. As climate change comes to almost wholly define the Pacific as a region, other pressing issues are either ignored or absorbed into the ways climate change is addressed, governed, and financed. However, PICTs are not passive in the ways climate change is constructed in the region and utilize the idea of the frontline, of shared fate, at the global level in international climate change negotiations. In the final section we focus on the role of PICTs in the AOSIS, and the possibilities for agency in global arenas. Without the powerful construction of the Pacific as the frontline of climate change, it is doubtful that states like Tuvalu or Kiribati would have any legitimate voice at the global level.

Chapter 4, 'Constructing Climate Security in the Pacific' takes one of the narratives discussed previously that has accelerated in its use and influence in the last decade, climate change as a security threat. Derived from the concept of environmental security, climate security has shaped understandings of climate change in the Pacific in the last decade. Climate security has also brought the future of the Pacific into discussions about regional stability, failed states, and refugee crises. These discussions reconfigure the Pacific as a risk, a region where climate-induced conflict within and between PICTs will collapse societies and institutions, affecting the stability of the greater regions, in particular Australia and New Zealand. However, climate security is not a singular narrative and climate security is also constructed as vulnerabilities that undermine human security. Building on existing literature we focus on six vulnerabilities: loss of livelihood, worsening poverty, food security, human health, gender, and migration. There are some overlaps with the general construction of climate change in the region, but here these vulnerabilities are linked explicitly to insecurity. While constructing climate security as vulnerabilities offers the most promise for addressing issues faced in the region, many actors, including governments in the Pacific, draw from both of these competing discourses where necessary.

The construction of the Pacific as the frontline of climate change creates particular understandings of the region, its needs, and the best ways to combat climate change. Chapter 5, 'Organizing Climate Finance in the Pacific' studies the architecture of climate finance in the Pacific, a subset of the global architecture. The demand for climate finance by Pacific states to combat climate change is derived from their inability to finance adaptation and mitigation projects from their own resources. However, as we demonstrate, climate finance is donor-driven and is entwined with development assistance, often using the same language, frameworks, norms, and relations of dependency. After examining the global architecture, we analyse the institutional framework for climate finance in the Pacific. We close by exploring the normative implications of an emerging climate finance regime in the Pacific and the impact of a fragmented, diverse, and complicated regime in the region where it matters most of all. We find that a combination of complex funding structures, inadequate sources of finance, and the absence of donor coordination reduce the effectiveness of climate finance in the Pacific.

The concluding chapter, 'The Future of Climate Politics in the Pacific', begins with the aftermath of the UNFCCC's COP 21 meeting in Paris in 2015. COP 21 was a crucial meeting for PICTs as an agreement that would set a commitment on a global temperature rise was on the agenda. PICTs pushed hard for a global commitment to a 1.5° C rise, a push framed, once again, as a matter of survival for PICTs. The resulting Paris Agreement fell short of expectations for Pacific states and the concluding chapter begins by analysing the implications for the Pacific following the Paris Agreement and what the future of regional climate politics may look like. The book closes with a summary of the main points and some thoughts on future research on climate politics at various scales in the Pacific.

Notes

1. The South Pacific Commission (SPC) was created in 1947 and although it changed its name to the Pacific Community in 1997 the acronym SPC has been retained because of widespread usage and recognition throughout the Pacific.

References

Farbotko, C. 2010. Wishful Sinking: Disappearing Islands, Climate Refugees and Cosmopolitan Experimentation. *Asia Pacific Viewpoint* 51 (1): 47–60.

Van Schendel, W. 2002. Geographies of Knowing, Geographies of Ignorance: Jumping Scale in Southeast Asia. *Environment and Planning D: Society and Space* 20 (6): 647–668.

CHAPTER 2

Organizing a Regional Response to Climate Change in the Pacific

Abstract The origins of contemporary regionalism in the Pacific are explored to understand the degree of regional interdependence, the strength of regional identity or regional awareness, and the creation of regional organizations that reflect and in some cases create this identity. The first part of the chapter provides a brief overview of regional governance, and analyses the sense of shared identity articulated as the 'Pacific Way'. Though difficult to define, the Pacific Way is the ideational underpinning of key regional organizations and a way of structuring decision-making in a manner that reinforces distinctiveness and a shared way of approaching cooperation and contention. The second part of the chapter maps the mandates, resources, and competencies of the various regional organizations concerned with climate change in the Pacific, illustrating both the complexity and coherence in the regional architecture.

Keywords Pacific regional governance • The Pacific Way • Pacific Islands Forum • Decolonization in the Pacific

The Pacific has a well-developed framework of regional organizations and institutions that deliver services in issue- areas such as security, economics, health, education, disaster management, and the environment. Delineating historical periods is necessarily problematic but we articulate two phases of Pacific regionalism: a colonial period and a post-independence or post-colonial

M. Williams, D. McDuie-Ra, *Combatting Climate Change in the Pacific*, https://doi.org/10.1007/978-3-319-69647-8_2

period. Arguably, post-independence regional cooperation in the Pacific has developed a distinctive dynamic based on the Pacific Way and grounded in the material realities of post-colonial political economies. Distinguishing between these two periods is useful but also limited. It is crucial to recognize the agency of Pacific leaders and communities in charting their own destiny rather than simply concluding that colonial arrangements were accepted in an uncritical and unreflective manner in the pre-independence period. Moreover, while we contend that two phases of regionalism can be identified, one cannot posit a stable set of relations in either period. And, within the post-independence period recent developments indicate important fractures in the stability of regionalism in the Pacific. Furthermore, there are elements of continuity as well as discontinuity between the two periods.

The history of regional intergovernmental cooperation in the Pacific can be dated to the colonial period. In 1947 the South Pacific Commission was created by the colonial powers (Australia, France, the Netherlands, New Zealand, the United Kingdom, and the United States) with the task of encouraging and strengthening 'international cooperation in promoting the economic and social welfare and advancement of the peoples of the non-self-governing territories in the South Pacific region'.[1] As various Pacific territories gained their independence, this existing framework provided a basis for further cooperation. Instead of abandoning the South Pacific Commission, newly independent Pacific states assumed membership and today all 22 island countries and territories of the Pacific are full members. However, as a reflection of these changing regional dynamics the South Pacific Commission changed its name to the Pacific Community (SPC) in 1997. The membership and funding of the SPC is indicative of one of the characteristics of Pacific regionalism. Unlike regional organizations in other developing regions, membership of Pacific regional organizations is frequently inclusive of colonial powers (in this case France and the United States) and the two dominant developed Pacific powers (Australia and New Zealand).

However, the functional mandate of the SPC failed to reflect the political priorities of newly independent states and the search for a regional organization with a more explicit political focus led to the creation of the Pacific Islands Forum (PIF). First convened in 1971, PIF is a political grouping of 18 independent and self-governing states in the Pacific.[2] PIF is the region's principal political and economic policy organization and Forum Leaders meet annually to develop collective responses to regional issues.

The SPC and the PIF are integral components of a patchwork of organizations that have emerged in the Pacific to provide regional solutions to common problems. Currently nine organizations are grouped together under the umbrella of the Council of Regional Organizations in the Pacific (CROP). These organizations are the Pacific Islands Forum Fisheries Agency (FFA), the Pacific Islands Development Programme (PIDP), the SPC, the PIF, the Secretariat of the Pacific Regional Environment Programme (SPREP), the South Pacific Tourism Organization (SPTO), the University of the South Pacific, the Pacific Power Association, and the Pacific Aviation Safety Office. CROP was established in 1998 'to improve cooperation, coordination, and collaboration among the various intergovernmental regional organizations to work towards achieving the common goal of sustainable development in the Pacific region'.[3] Environmental issues feature prominently on the agendas of three CROP organizations: the PIF, the SPC, and the SPREP. The three organizations have each developed separate but at times overlapping mandates in relation to climate governance.

Regionalism in the Pacific is shaped by specific material and normative factors. The material framework of regional cooperation is the political economy of the region. As small isolated states with scarce resources, 'narrow' economies, weak governance and infrastructure, and protracted economic growth, the region's constraints and challenges are important common factors that bind it together (Chand 2010: 12; Herr and Bergin 2011: 11). Economic growth, sustainable development, security, governance, and climate change dominate discourses of Pacific regional cooperation and have shaped its trajectory (ADB 2005: xiv; PIFS 2010; UNESCAP 2013: 4). By operating collectively Pacific Island states can have a stronger presence in international trade negotiations as well as have a better chance to access the international market (Frazer and Bryant-Tokalau 2006: 2).

The economies of Pacific Island Countries and Territories (PICTs) rely heavily on tourism (Harrison and Prasad 2013), as well as on overseas remittances (Browne and Mineshima 2007; Makun 2017). And this reliance is subject to the vagaries of the global economy. For example, receipts from both tourism and remittances were slowed down by the global financial crisis (UNESCAP 2013: 1). Despite the absence of reliable comparative data it is generally accepted that unemployment is rife in the majority of island states, especially amongst young people (Barbara and McMahon 2016), and the agricultural sector remains at the subsistence level

(UNESCAP 2013: 1). Land resources are scarce and, further, the ratio of land to sea is vastly disproportionate. Tuvalu, for instance, comprises 26 km^2 distributed over nine islands (Herr and Bergin 2011: 10). Marine resources are therefore essential to the region's economic possibilities (Herr and Bergin 2011: 10). The Pacific zone is one of the richest for tuna fishing (Chand 2010: 13). To this end, the FFA, founded in 1979, aimed to secure and manage fisheries in the Pacific by establishing 200-mile exclusive economic zones. This initiative has been one of the 'most successful' displays of the benefits of regional Pacific initiatives (Frazer and Bryant-Tokalau 2006: 8). It has been a major economic contributor to the region, providing an average of 7% of each member country's GDP (Herr and Bergin 2011: 18; ADB 2005: 66). The exclusive economic zones generate revenue from access fees paid by foreign vessels, fish processing and exports, national fleet catch and harvesting, and employment. However, even this success unveils many of the challenges faced in the region. Despite contributions to GDP for members (which peaked in 2012), the FFA points out that the costs of fishing industries in PICTs are much higher than competitors in Asia given the higher cost of electricity, fuel, and labour and they estimate the total cost per tonne processed in FFA member countries is 3.7 times that in Asia (FFA 2017: 12).

According to the Asian Development Bank (ADB) (2005: 1), resource endowment and geography, which initially gave shape to regional strategies, will continue to define regionalism in the Pacific. However, concerns with national sovereignty rather than regional interests continue to challenge an integrated, cohesive regional approach (Frazer and Bryant-Tokalau 2006: 4–5; Chand 2010: 12–17). Additionally, the same reasons for regional cooperation are also the primary challenges to regionalism, namely, the fact that the majority of states comprise a vast number of small, culturally diverse islands with fragile governing structures (Chand 2010: 1; Herr and Bergin 2011: 8).

The position of Australia and New Zealand in the Pacific is an important element in the 'disjointed' nature of the Pacific region. Although both powers belong to the region, their involvement is not based on common regional goals, but rather, as the primary donors. Moreover, they have been viewed as shaping the direction of regional strategies according to their own interests (Frazer and Bryant-Tokalau 2006: 2; Herr and Bergin 2011: 12). Security concerns of external powers, especially Australia and New Zealand have also played an important role in shaping the current Pacific regional structures (Chand 2010: 8). The absence of a perceived sense of 'nationality' in many of the larger states, such as the Solomon Islands, and the threat of consequent civil unrest, have attracted—and still depend on—external assistance and intervention (Herr and Bergin 2011: 11). It can be argued that the geographic features of the Pacific region further prevent a coherent regional strategy (Herr and Bergin 2011: 13). Their distance from 'major global centres' as well as from each other, in addition to the small size of the islands and to infrastructural and technological limitations, constrain cooperation (Herr and Bergin 2011: 13). Further, only three members of the PIF have a military body, while the remaining 11 members largely rely 'on the absence of external threats and the protection of the international system for their security' (Herr and Bergin 2011: 12).

The ill-defined and poorly conceptualized concept of the Pacific Way has provided the normative basis of post-independence regionalism in the Pacific. The term 'Pacific Way' was coined by the then prime minister of Fiji, Ratu Sir Kamisese Mara at the UN General Assembly (UNGA) in October 1970. He used it in reference to the independence processes in the Pacific that were marked by relatively peaceful transitions from colonial rule in comparison to other geographic regions. He noted in reference to the independence of Fiji,

> Many speakers have commented on our peaceful transition to independence. But this is nothing new in the Pacific. Similar calm and orderly moves to independence have taken place in Western Samoa, the Cook Islands, in Nauru, and in Tonga. We like to think that this is the Pacific Way, both geographically and ideologically. As far as we are authorized by our friends and neighbours, and we do not arrogate to ourselves any role of leadership, we would hope to act as representative and interpreter of that voice. (Mara 1997: 238)

Since this formative speech, the Pacific Way has been interpreted to apply to the South Pacific region and has become embedded as the ideational underpinning of many of its regional organizations. Nevertheless, the exact and precise definition of the Pacific Way remains elusive. Ratu Mara failed to elaborate on the term in subsequent speeches or writings. Despite the absence of a consensus definition most analysts use the term to refer to an approach to conflict resolution, and a negotiating tool based on consensus and compromise (see, e.g. Crocombe 1976; Lawson 2010; Mishra 2005; Rolfe: 2001). It is now generally accepted that the Pacific Way is a post-colonial concept (Mishra 2005) that is used to unify the

PICTs and to assert the Pacific's control over regional decision-making. As a basis for regional cooperation the Pacific Way has been adopted by the PIF, formed one year after Ratu Mara's speech to the UNGA. For example, the Pacific Islands Forum Secretariat (PIFS) notes that their guiding principle and values are rooted in 'the cultural diversity of the region with tolerance and respect (The Pacific Way)' (PIFS, n.d). Leading officials have made explicit reference to the Pacific Way as the basis for cooperation in the region. For example, on the occasion of the 40th PIF leaders meeting in 2009 (held in Cairns, Australia), Tuiloma Neroni Slade the Secretary General of the PIFS stated: '[t]he experience of the Forum in our past 39 meetings, points to the inevitability of togetherness. The Pacific is at its best when it acts as a region. In times of crisis it is the natural way. It is the very essence of the Pacific Way' (Slade 2009). And Henry Puna, prime minister of the Cook Islands when commenting on Fiji's return to the African, Caribbean and Pacific meetings in 2013, stated: 'The Pacific came together as a family and dealt with an important issue in a way that a family should- a Pacific Way' (Komai 2013). Furthermore, the term has been incorporated into key documents. For example, the 2003/2004 Annual Report of the PIFS claims that the Pacific Way is a 'philosophy that guides the region's approach to political, economic and social issues' (PIFS 2003/2004: 2).

However, the extent to which the Pacific Way provides a normative foundation for a post-colonial regionalism in the Pacific has been questioned by Lawson who argues that in its original formulation the Pacific Way was not framed as a post-colonial discourse but rather as a conservative discourse framing common interests of the elite social groups among the colonized and colonizers (Lawson 2010: 310). She contends that the Pacific Way obtained its post-colonial status in later years. Lawson's critique has resonance but from our perspective it is not the origins of the term that is significant but rather contemporary interpretations. We can accept Lawson's critique of the origins of the term without undermining the argument made here since reference to the Pacific Way continues to be used to support post-colonial discourses of regionalism.

Rhetorical reference to the Pacific Way is reflected in the decision-making structures of key regional organizations. The PIF does not have a formal voting structure, and decisions are made by consensus. In other words, the emphasis is on resolving disputes in the Pacific Way. And while the SPC, has a formal voting system this is rarely resorted to since it is now widely accepted that debates are 'usually resolved in the Pacific way by consensus'.[4] Consensus decision-making systems are frequently criticized from the perspective of efficiency. And the use of the Pacific Way as the normative basis of decision-making in the SPC and the PIF has been sharply criticized for creating stasis in policy-making, delaying the resolution of issues, and diluting the strength of Forum Communiqués (Shibuya 2009: 114). Moreover, the absence of voting need not indicate a causal link between the Pacific Way as a normative foundation and the outcome of decisions. The unequal distribution of power in the region can create the appearance of consensus, thus masking the structural and institutional power of Australia and New Zealand, the two dominant industrial countries in the region. Pacific leaders are cognizant of the power differential and the influence accruing to Australia and New Zealand as a function of their financial contributions to regional organizations. And attempts by Australia and New Zealand to speed up negotiations such as those on a regional trade agreement have led some observers to question the inclusion of these states within the decision-making framework in the region (Baker 2015). Noel Levi, then Secretary General of the Forum Secretariat, captured this friction between the impatience of the metropolitan powers and the consensus approach of the South Pacific states and their commitment to the Pacific Way when he claimed that while other states from the outside want to come to a decision, 'here in the Pacific, we take our time' (Shibuya 2009: 114).

The 'failure' of the Pacific Plan and a more assertive role by Fiji has led to a degree of turbulence in regional arrangements in the Pacific. These developments indicate a fracturing of the post-independence consensus and signal the transition to a form of more assertive form of post-colonial regionalism. As Tarte (2014: 312) writes, 'This new dynamism is driven by the discontent of a growing number of island states with the established regional order identified by prevailing institutions, power, and ideas, and by a desire to assert greater control over their own futures'.

To some extent this post-colonial regionalism challenges the pan-Pacific idealism of the post-independence approach to regionalism. For example, it is worth noting that in recent years several sub-regional and extra-regional groupings have emerged, including the Melanesian Spearhead Group, which has been successful in creating its own free trade area (see Marawa 2015). There have been initiatives that connect states in the Pacific and Southeast Asia, such as the Coral Triangle Initiative that seeks cooperation on marine conservation and fisheries between Indonesia, Malaysia, Timor-Leste, the Philippines, the Solomon Islands, and Papua

New Guinea (Rosen and Olsson 2013). The Coral Triangle Initiative receives funds through some of the same mechanisms as Pacific organizations and also partners with SPREP. However, it only has two Pacific members and is peripheral compared to the Pacific Way solidarity, and climate-related governance bounds the region. Other groups based on former colonial and shared linguistic and cultural elements also come into play from time to time on certain issues; however, climate change is less prominent at this scale and thus we have not focused on these in detail.

Pacific Regional Organizations: Organizational Structure and Political Process

The second section of the chapter identifies the mandates, resources, and competencies of three key regional organizations concerned with climate change in the Pacific.

The Pacific Islands Forum

The PIF represents the highest political organ of Pacific regionalism. Created in 1971 full membership is restricted to independent states. The 18 member states of the PIF are Australia, Cook Islands, Federated States of Micronesia, Fiji, French Polynesia, Kiribati, Nauru, New Caledonia, New Zealand, Niue, Palau, Papua New Guinea, Republic of Marshall Islands, Samoa, Solomon Islands, Tonga, Tuvalu, and Vanuatu. Associate member status has been granted to Tokelau (2014). Special observer status has been conferred on Wallis and Futuna (2006), the Commonwealth (2006), the United Nations (2006), the ADB (2006), Western and Central Pacific Fisheries Commission (2007), the World Bank (2010), the African, Caribbean and Pacific Group of States (ACP) (2011), American Samoa (2011), Guam (2011) and the Commonwealth of the Northern Marianas (2011), Timor-Leste (2002), and the International Organization for Migration (2014).

The PIF is an intergovernmental organization and the highest decision-making body is the annual Forum meeting chaired by the head of government of the host country. The transition from a leaders' meeting to an international organization was achieved with the 2005 Agreement Establishing the Pacific Islands Forum, which provided the PIF with a legal basis. There is no formal constitution, which means any topic can be discussed and there are no formal voting structures to encourage decision-making by consensus (Shibuya 2009: 105). This approach to decision-making reflects the normative foundation of the PIF. The Pacific Way emphasizing consensus and solidarity has been endorsed as the cornerstone of the organization's value system. The leaders of the member countries meet annually to respond to regional issues. The decisions made are reflected through Forum Communiqués that establish the position taken by the group on issues such as climate change, nuclear testing, fisheries, and security/environmental issues. The Communiqués often note the problems, progress, or lack of progress on these issues rather than provide programmes to solve them. In addition to the leaders' meeting, a Forum Regional Security Committee is also convened on an annual basis and is attended by officials from member countries and representatives of the Regional Law Enforcement Secretariats, as well as representatives from other CROP agencies. The PIF is generally regarded 'as the authentic policymaker for the region as a whole' (Herr and Bergin 2011: 15) as for the past 20 years it is has played a significant role in bringing regional issues, especially climate change, to the international arena although there are calls for the PIF to address island issues more forcefully away from the interests of Australia and New Zealand (Tavola 2015).

The PIFS based in Suva, Fiji, provides the continuity to the PIF. The PIFS conducts the day-to-day operations of the Forum and implements the decisions taken at the annual meetings in furtherance of the PIF's primary goal of stimulating economic growth and enhancing political governance and security for the region. The secretariat is tasked with three key functions: (a) providing policy advice and guidance in implementing the decisions of the Leaders Forum; (b) coordinating and assisting in the implementation of decisions; and (c) supporting the leaders' meetings, ministerial meetings, and the various committees and working groups. The PIFS is a relatively small secretariat with 127 staff, mainly based in Suva.

A Secretary General appointed for a three-year term by Forum Leaders heads the PIFS. In the absence of a management board or executive committee the Secretary General is directly responsible to Forum Leaders and to the Forum Officials' Committee. The Secretary General is also the permanent Chair of the CROP. The current Secretary General of the PIFS is Dame Meg Taylor from Papua New Guinea who assumed office in December 2014 (see Table 2.1 for a list of PIFS Secretaries General). The Secretary General is assisted by an Executive Management team comprising of two Deputy Secretaries General and four Programme Directors.

Table 2.1 Secretaries General of the PIFS

Name	*Nationality*	*Period in Office*
Henry Naisali	Tuvalu	1988–1992
Ieremia Tabal	Kiribati	1992–1998
Noel Levi	Papua New Guinea	1998–2004
Greg Urwin	Australia	2004–2008
Tuiloma Neroni Slade	Samoa	2008–2014
Meg Taylor	Papua New Guinea	2014–

The members of this team are responsible for managing their respective programmes and collectively provide leadership to the rest of the organization.

The PIFS's activities are financed through a budget made up of three components—a regular budget derived from member contributions based on economic size, a core budget based on voluntary contributions from members, and an extra budget derived from funding provided by non-members. For instance, in 2013 the annual budget (in Fiji dollars) was constructed as follows: $8.6 million (regular budget), $14 million (core budget), and $13.9 million (extra budget). PIFS finances are heavily dependent on Australia and New Zealand. In 2013, Australia contributed $16.1 million to the total budget and New Zealand contributed $4 million. The entire core budget was funded by these two states.

The PIFS's role in regional governance is centred on three strategic programmes: the Economic Governance Programme, the Political Governance and Security Programme, and the Strategic Partnerships and Coordination Programme. These programmes provide policy advice and expertise to member states and works with other agencies to provide assistance in these areas.

Between 2005 and 2013 the Pacific Plan whose main aim was to strengthen 'regional cooperation and integration' (PIFS 2013: 11) formed the basis of the PIFS work programme. The Pacific Plan was formulated by the PIF in 2005 and at its inception it was considered a 'living document' outlining areas of work in the following categories: economic growth, sustainable development, good governance, and security. The Plan arose in response to a request from the Chair of the PIF, then New Zealand Prime Minister, Helen Clark, to carry out a review of the Forum's role, its functions, and the Secretariat. It was the first comprehensive review of the Forum since its inception. The stated aim of the Pacific Plan was to strengthen regional cooperation and integration across a broad range of governance objectives, particularly focused on 'economic growth, sustainable development, good governance and security' (PIFS 2005a). The Plan outlined a particular approach to regionalism: '[r]egionalism under the Pacific Plan does not simply any limitation on national sovereignty. It is not intended to replace any national programmes, only to support and complement them. A regional approach should be taken only if it adds value to national efforts' (PIF 2005b: 3). A comprehensive review of the Pacific Plan was undertaken in 2013 and the Framework for Pacific Regionalism endorsed by Forum Leaders in July 2014, provides the current framework through which to address strategic issues in the region.

Adopted by Pacific leaders in 2014 the Framework for Pacific Regionalism established a basis for a deeper regionalism. In this context regionalism was defined as 'The expression of a common sense of identity and purpose, leading progressively to the sharing of institutions, resources, and markets, with the purpose of complementing national efforts, overcoming common constraints, and enhancing sustainable and inclusive development within Pacific countries and territories and for the Pacific region as a whole' (PIFS 2014: 1).

The Framework for Pacific Regionalism like its predecessor, the Pacific Plan, establishes a normative basis for regionalism in the Pacific. It represents a commitment to a deeper level of integration and establishes a process for developing strategic objectives, and processes for the setting of priorities.

PIFS and Climate Governance

The PIF has played two important roles in climate governance. First, it has been instrumental in setting the framework for overall regional governance. As such it addresses many of the institutional deficiencies inhibiting cooperation broadly, and is able to channel financial resources into regional cooperation. In the context of climate change the annual meeting of PIF leaders has provided a platform for bold political declarations. Climate change has featured in every PIF communiqué since 1998. In 1998 the Communiqué issued after the 19th PIF 'expressed concern about climatic changes in the South Pacific and their potential for serious social and economic disruption in countries of the region' (PIF 1988). The PIF communiqués have confirmed the various activities and regional frameworks developed relating to climate change and point to ongoing discussions through the Forum about how to make them function well together. The Forum established the current principles underlying collective action and the strategic framework of regional priorities through three major documents. In October 2000 Forum Leaders adopted the *Pacific Islands Framework for Action on Climate Change, Climate Variability, and Sea Level Rise, 2000–2004*. This plan was updated in 2005 when Forum Leaders endorsed the *Pacific Islands Framework for Action on Climate Change* (PIFACC) as a regional mechanism to support responses to climate change through to 2015. The Framework outlined the priority action areas for the Pacific and its central objective was to ensure that Pacific island peoples and communities build their capacities to be resilient to the risks and impacts of climate change. Seven principles shaped the attainment of the key objectives: implementing adaptation measures; governance and decision-making; improving understanding of climate change; education, training, and awareness; contributing to global greenhouse gas reduction; and partnerships and cooperation. The Framework was supplemented in 2007 when the Pacific island governments adopted an action plan to carry out the PIFACC, in which national activities are complemented by regional programming. Further, in 2008 the Pacific Forum meeting in Niue adopted the *Niue Declaration on Climate Change* (PIF 2008). The Niue Declaration is the principal political climate change statement of the Pacific region. It calls for urgent action by the world's major greenhouse gas emitting countries to set targets and make commitments to significantly reduce their emissions, and to support the most vulnerable countries to adapt to and address the impacts of climate change.

The experience with the PIFACC and the continued salience of climate change as a political issue necessitated the creation of a successor framework when the PIFACC expired in 2015. The current framework for cooperation on climate change was endorsed by Pacific Island Forum Leaders, at their 47th meeting in Pohnpei in 2016. The Framework for Resilient Development in the Pacific: An Integrated Approach to Address Climate Change and Disaster Risk Management (FRDP) was the outcome of a process initiated by Forum Leaders in 2012. A steering committee made up of representatives from the SPREP, the Pacific Community Committee of Representatives of Governments and Administrations, the Regional Disaster Managers' Meeting, the Pacific Climate Change Roundtable, the Pacific Meteorological Council, the Forum Economic Ministers Meeting, the French Territories, the Pacific Islands Alliance of NGOs, and the Pacific Islands Private Sector Organisation held extensive consultations with stakeholders at the local, national, regional, and international levels. The Framework for Resilient Development merges the concerns of two previous regional framework: PIFACC and the Pacific Disaster Risk Reduction and Disaster Management Framework for Action. The new integrated framework recognizes the interconnection between climate change and disaster risk management.

The FRDP is mandated until 2030. It is based on three main goals:

1. Strengthened integrated adaptation and risk reduction to enhance resilience to climate change and disasters
2. Low-carbon development
3. Strengthened disaster preparedness, response, and recovery

The FRDP marks an evolution in regional consultation on climate change. Unlike its predecessors as a regional framework, the FRDP presents an integrated framework that goes beyond identifying climate change risks and vulnerabilities to linking these with disaster risk reduction. The FRDP recognizes the differential ability of PICTs to respond to climate-induced vulnerability. It therefore embraces a voluntary approach to action and resists a prescriptive solution.

The PIF has welcomed various climate finance pledges from bilateral donors including Australia, Japan, and the European Union. The salience of climate change was highlighted when it was listed as an area for attention in the ongoing implementation of the Pacific Plan in 2008. In support of these efforts PIF tasked SPREP to urgently undertake a review of the region's meteorological services. It called on SPREP, SPC, SOPAC, and the University of the South Pacific to work towards implementing regional actions on climate including to 'collaboratively to rationalize the roles of the various regional organizations and to harmonize donor engagement' (PIF 2008: 13).

Second, the PIF has supported the external activities of the Pacific states in international negotiations. This has improved the knowledge gaps on climate change among PIF members and also brought knowledge of the climate change scenario in the Pacific to an international audience. Since 1993, the PIF has frequently endorsed the representations of the Association of Small Island States (AOSIS) at the United Nations Framework Convention on Climate Change (UNFCCC) and other international negotiations. For example, following the *Global Conference on the Sustainable Development of Small Island Developing States* held in Barbados

(1994), the PIF decided to 'establish a regional consultative mechanism to coordinate and facilitate implementation of the Barbados Conference's outcomes as recommended in paragraph 132 of the Program of Action' (PIF 1994). And prior to the UNFCCC Conference of the Parties (COP) held in Copenhagen in 2009 the PIF stated that

> For Pacific Island states, climate change is the great challenge of our time. It threatens not only our livelihoods and living standards, but the very viability of some of our communities. Though the role of Pacific Island States in the causes of climate change is small, the impact on them is great. Many Pacific people face new challenges in access to water. The security of our communities and the health of populations is placed in greater jeopardy. And some habitats and island states face obliteration. (PIF 2009)

Thus, it can be said that the PIF plays a role in responses to climate change within the region and in bringing the challenges faced by the region to a global audience.

The Pacific Community

Within the fabric of Pacific regionalism the Pacific Community is the oldest organization. Founded in 1947 as the South Pacific Commission the organization attained its current name on its 50th anniversary in 1997. In contradistinction to the PIF the SPC was founded as a non-political body. The current membership of the SPC stands at 26 countries thus making it the largest regional organization in the Pacific. The governing body of the SPC is the biennial Conference of the Pacific Community. Each member state has one vote although voting is rare with the Pacific Way of making decisions by consensus the dominant mode of arriving at decisions in the organization. The Conference of the Pacific Community is complemented by the Committee of Representatives of Governments and Administrations, which meets annually and is empowered to make decisions on the governance of SPC in the years that the Conference of the Pacific Community does not meet.

The principal activity of the SPC is to undertake research and provide technical assistance and training in support of the economic and social development of its 26 members. The SPC's work programme is mapped into six divisions: the Economic Development, Fisheries, Aquaculture and Marine Ecosystems Division, Geoscience, Land Resources, Public Health,

Table 2.2 Directors General of the Pacific Community

Name	*Nationality*	*Period in Office*
Ati George Sokomanu	Vanuatu	1993–1996
Robert B. Dun	Australia	1996–2000
Lourdes T. Pangelinan	Guam	2000–2006
Jimmie Rodgers	Solomon Islands	2006–2014
Colin Tukuitonga	Guam	2014–

and the Statistics for Development Division. The secretariat houses four work programmes: Climate Change and Environmental Sustainability, Educational Quality and Assessment, Regional Rights Resource Team, and the Social Development Programme.

A Director General who is assisted by two Deputy Directors General heads the SPC. The current Director General is Dr Colin Tukuitonga from Niue, who assumed office in January 2014 (see Table 2.2 for previous Directors General). The headquarters of the SPC is in Nouméa but six organizational units (the Economic Development Division, the Educational Quality and Assessment Team, the Geoscience Division, the Land Resources Division, the Regional Rights Resource Team, and the Social Development Team) are based in solely Suva. The organization also has a Melanesia Regional Office in Port Vila, Vanuatu, and a Micronesia Regional Office in Pohnpei, Federated States of Micronesia.

Funding for SPC activities is derived from members and non-members. Funding is provided under two budget headings: core and non-core. Core funding is derived solely from member countries and non-member countries provide non-core funding. In 2015 the total income[5] of the SPC was $116,473,775, of which $36,490,628 was provided by members and $79,983,147 by non-members. Members contributed the majority of their funding as a core contribution. The metropolitan members made the highest member contributions. Australia contributed the most to the SPC, $23,239,658 (27.01%), followed by New Zealand ($4,677,009), France ($2,623,877), and the United States ($2,536,508), which makes up approximately 93% of the total member government contributions made to the SPC. In terms of non-members, the two highest financial contributions to the SPC were from the European Union ($25,754,986), around 29.93% of total funding (see Pacific Community 2016: 27).

SPC and Climate Governance

The SPC is involved in attracting finance for climate change-related projects in member states and territories. It has also been instrumental in building knowledge across a number of development-related fields in the region. Indeed, the SPC is the Pacific's largest development organization, and provides technical assistance, policy advice, training, and research services to members in areas such as health, human development, agriculture, forestry, and fisheries. SPC has positioned itself as the lead agency for a range of climate-related activities and has secured funding from external agencies for capacity building, land-based, and coastal adaptation measures. Among its various projects five are noteworthy and demonstrate the breadth of its activities and the scope for overlap between the three organizations with an environmental mandate.

The SPC was the lead agency for a European Union-funded project, the objective of which is to assist PICTs in developing detailed climate change response strategies and investment plans and to integrate these into consistent overarching national climate change response frameworks (SPC 2011). The project titled *Increasing Climate Resilience of Pacific Small Islands States through the Global Climate Change Alliance* was operational in the Cook Islands, Kiribati, Marshall Islands, Federated States of Micronesia, Nauru, Niue, Palau, Tonga, and Tuvalu. Through its Coast Fisheries Programme, the SPC convenes a climate project funded by AusAid titled the *Vulnerability and Adaptation of Coastal Fisheries to Climate Change*. The aim of the project is to design and field test pilot projects to determine whether changes are occurring in the productivity of coastal fisheries and, if changes are found, to identify the extent to which such changes are due to climate change as opposed to other causes. The Land Resources Division of SPC convened the (2009–2015) regional programme *Coping with Climate Change in the Pacific Island Region*. The programme, which began with pilot projects in Fiji, Tonga, and Vanuatu, has been expanded to include projects in Federated States of Micronesia, Kiribati, Marshall Islands, Nauru, Palau, Papua New Guinea, Samoa, Solomon Islands, and Tuvalu. The initiative for this programme emerged from the 2006 meeting of the Heads of Agriculture and Forestry Services (HOAFS). The initial focus of this programme was on land-based natural resources such as agriculture, forestry, and land use. Over its lifespan it shifted focus to a range of measures spanning adaptation and mitigation including energy projects, tourism PPPs, education, and REDD. SPC is also the lead agency for REDD pilot activities (2009–2014) in Fiji, Papua New Guinea, Solomon

Islands, and Vanuatu. The project titled *Climate Change Protection through Forest Conservation in Pacific Island States* had three components: creation of a Pacific Regional REDD+ Policy Framework, creation of a REDD+ information and support platform, and development of REDD+ readiness at a national level. Funded under the auspices of the German International Climate Initiative the project had a budget of €4.9 million (SPC LRD 2011; Federal Republic of Germany 2016)

Between 2012 and 2015 in collaboration with United States Agency for International Development (USAID), the Pacific Community implemented the USAID/SPC Food Security Project. Recognizing the impact of climate change on long-term food security in the region the project focused on food production systems in Fiji, Kiribati, Samoa, the Solomon Islands, Tonga, and Vanuatu. Its aims were inter alia to improve understanding of the impact of climate change on food security and to develop coping strategies.

Secretariat of the Pacific Regional Environmental Programme

The creation of SPREP was the outcome of increasing awareness of environmental issues in the region, and recognition that the existing framework was unable to fulfil the needs of PICT governments. However, the path from this new environmental consciousness to the creation of an independent regional organization was a slow and tortuous one subject to the politics of the region and perhaps best traces the trajectory of climate change from a fringe concern to centrality. In 1973 a regional ecological adviser was appointed to the SPC. By 1978, SPC, the United Nations Environment Programme (UNEP), the South Pacific Bureau for Economic Co-operation (SPEC, now the PIFS), and the Economic and Social Commission for Asia and the Pacific (UNESCAP) had decided that a comprehensive environment programme for the region was a necessity. However, it was not until 1982, at a Conference on the Human Environment in the South Pacific, held in Rarotonga, Cook Islands, that participants decided to establish SPREP. At the outset, SPREP was created as a separate entity within SPC and based in Nouméa. SPREP became an autonomous intergovernmental organization with the signing of the Agreement Establishing SPREP in Apia on 16 June 1993.

SPREP is the principal regional environmental organization and its role is to promote cooperation in the South Pacific Region and to provide assistance in order to protect and improve the environment and to ensure sustainable development for present and future generations. In pursuit of securing its mandate SPREP is involved in knowledge creation and sharing, maintaining an institutional 'home' that acts as an arena for deliberation and decision-making, and acting as an interface between donors and local partners throughout the region. The member countries of SPREP are American Samoa, Australia, Cook Islands, Federated States of Micronesia, Fiji Islands, France, French Polynesia, Guam, Kiribati, Marshall Islands, Nauru, New Caledonia, New Zealand, Niue, Northern Mariana Islands, Palau, Papua New Guinea, Pitcairn Islands, Samoa, Solomon Islands, Tokelau, Tonga, Tuvalu, the United States, Vanuatu, and Wallis and Futuna.

SPREP is bound by the *Agreement Establishing the South Pacific Regional Environment Programme* and its 2004 Amendment. The organs of SPREP are the SPREP Meeting and the Secretariat. The highest forum in SPREP is the Annual Meeting. SPREP is also the secretariat for two regional conventions[6]: the Convention for the Protection of the Natural Resources and Environment of the South Pacific Region and related protocols (Nouméa Convention) and the Convention to Ban the Importation into Forum Island Countries of Hazardous and Radioactive Wastes and to Control the Transboundary Movement and the Management of Hazardous Wastes within the South Pacific Region (Waigani Convention). The annual meetings of the Nouméa Convention and the Waigani Convention are held at the same time as the Annual SPREP Meeting. In common with other regional organizations in the region decisions made in the SPREP Meetings are based on consensus of the parties.

SPREP is headed by a Director General who oversees the work of the organization headquartered in Apia, Samoa. The current Director General, Kosi Latu, is the seventh head of the organization (see Table 2.3). SPREP's current work programme is organized into four strategic priority areas: Biodiversity and Ecosystem Management, Climate Change, Environmental

Table 2.3 Executive Heads of SPREP

Name	*Nationality*	*Period in Office*
Tusani Joe Reti	Samoa	1981–1989
Vili Fuavao	Tonga	1990–1996
Tamari'i Tutangata	Cook Islands	1998–2002
Asterio Takesy	Federated States of Micronesia	2003–2008
David Sheppard	Australia	2009–2015
Kosi Latu	Samoa	2016–

Monitoring and Governance, and Waste Management and Pollution Control. A director who oversees the programmes within each area heads each strategic priority area.

The SPREP is funded through core contributions by its member states and income derived from its management services, and programme contributions that are voluntary in nature and tied to specific projects. In 2014 its total income of $19,272,903[7] was made up of $ 1,342,600 in core contributions, and $15,650,858 in programme contributions. Major bilateral funders were Australia, France, New Zealand, the United Kingdom, and the United States of America accounting for approximately 71% of total members' contributions. Multilateral funders contributed the bulk of programme funding (SPREP 2014: 58–59).

SPREP and Climate Governance

Climate change is central to the work of SPREP. It is enshrined in its work programme as Priority 1 and the organization was tasked with meeting the following goal:

> By 2015, all Members will have strengthened capacity to respond to climate change through policy improvement, implementation of practical adaptation measures, enhancing ecosystem resilience to the impacts of climate change, and implementing initiatives aimed at achieving low-carbon development.[8]

SPREP is thus specifically concerned with assisting PICTs in the implementation of adaptation and mitigation priorities and does so through two important roles. First, it coordinates regional activities. In support of this role SPREP has convened annual Climate Change Roundtables since the first was held in Apia, Samoa, in October 2008. The main objective of these meetings is 'to facilitate regional coordination, collaboration and coherence in the implementations of PICTs climate change priorities consistent with the Pacific Islands Framework for Action on Climate Change (PIFACC)'. In 2016, 134 representatives from all major intergovernmental organizations, regional organizations, and national government agencies were in attendance.

SPREP's second role is to provide assistance to PICTs in environmental protection and the implementation of sustainable development. In pursuit of this role SPREP has initiated important projects relating to climate change. We highlight three key projects that have recently been completed. The Global Environment Facility (GEF)- funded Pacific Islands Greenhouse Gas Abatement through Renewable Energy Project (PIGGAREP) was started in 2006; and came to an effective conclusion in 2014. The aim of the project was to reduce the growth of fossil fuel use in the Pacific region by 33%; and preliminary evaluations suggest that it achieved a 30% reduction compared to a Business As Usual prediction (SPREP 2017). The project was implemented in 11 PICTs—the Cook Islands, Fiji, Kiribati, Nauru, Niue, Papua New Guinea, Samoa, Solomon Island, Tonga, Tuvalu, and Vanuatu with UNDP, the GEF, and SPREP as implementing partners. The GEF provided a US$5.25 million project grant from the GEF Trust Fund (climate change focal area) with US$20.88 million in co-financing from PICT governments (approx. US$16 million) small amounts from SPREP and local and development banks have been approached (US$1 million).[9]

In 2009, SPREP commenced the *Pacific Mangroves Initiative* (2009–2014) in Fiji, Vanuatu, Solomon Islands, Samoa, and Tonga. The programme was funded by the German Federal Ministry for Economic Cooperation and Development (BMZ) (US$3.44 million disbursed), and administered by the International Union for the Conservation of Nature (IUCN), WorldFish Center, Solomon Islands, and the Institute of Applied Science at the University of the South Pacific. SPREP also hosted the Pacific Adaptation to Climate Change Project (2009–2014) implemented in the Cook Islands, Fiji, Micronesia, Marshall Islands, Nauru, Niue, Papua New Guinea, Palau, Solomon Islands, Tonga, Tuvalu, Vanuatu, and Samoa. The UNFCCC Special Climate Change Fund funded the project with US$13.13 million disbursed to date.

Constraints on Effective Regional Climate Governance

A regional framework for climate governance has emerged in the Pacific. Island governments have recognized that climate change is a priority issue and that national capacity is ill-equipped to respond to the challenges posed by climate variability. Adaptation measures and mitigation efforts are designed at the national level but supported by a regional framework. Adaptation measures include preparing coastal areas and infrastructure for more severe conditions, disaster risk management, improved agricultural methods, crops, fisheries, and forest management. Mitigation efforts have been directed at reducing fossil fuel use, improving energy efficiency, developing renewable energy (such as wind and solar power), and reducing green house gas emissions from forest degradation and deforestation. Despite the emergence of this regional framework, three constraints on effective regional cooperation can be delineated.

First, the stimulus for regional cooperation arises from the common threats and vulnerabilities presented by climate change and resource constraints in the PICTs. PICTs generally lack the qualified staff and institutions needed for climate change–related research and for programme development and implementation. Developing cooperation at the regional level is one response to the limited resources of the PICTs. However, the regional solution is itself heavily dependent on external financing. PIF, SPC, and SPREP are all reliant on external partners to provide financial support for adaptation and mitigation efforts. Many development partners have contributed significant funding for regional initiatives. Key funders include regional powers—Australia and New Zealand—and the European Union and Japan. Furthermore, with human resource limitations in PICTs, exacerbated by the brain drain and elevation of the remittance economy, key staff in regional organizations are often nationals of donor countries rather than from the PICTs. Our interviewees highlighted the ways in which dependence is a potential constraint on the direction of regional environmental governance in three ways. First, priorities can be shaped by the dictates of the funding agencies and conflicts may arise over specific adaptation or mitigation projects and priorities. Second, reductions in funding can result in the cancellation of projects deemed important by the region. Similarly, the movement of professionals in and out of key organizational roles makes it very difficult for regional organizations to retain human resources and organizational knowledge. Third, inflexibility and bureaucracy of donor programmes can create coordination problems for regional organizations as they negotiate with bilateral donors who all have different conditions and processes for the access and delivery of funds.

A second constraint arises from the political economy of the PICTs. Most countries in the region are not only poor but their economies are dependent on a limited number of sectors for foreign exchange and investment as will be discussed in the following chapter. The major industries in the regions such as tourism, fisheries, forestry, mining, and agriculture are all highly susceptible to external shocks. The global recession since 2008 has affected PICTs mainly as a result of a fall in demand for commodity exports, declining tourism receipts, and reductions in earnings from remittances. This economic downturn is being experienced at the household, local, and national levels. This increased economic vulnerability affects regional environmental cooperation since it tends to turn attention away from the environment and towards economic development activities. Of course, the official discourse of sustainable development theoretically combines environmental and development activities but in reality these issues tend to be in competition for resources. There is an additional complication here. The construction of climate change as a threat makes it difficult for national and international actors, both state and non-state, to effectively politicize other environmental issues. Environmental issues that challenge core aspects of PICT economies are even harder to politicize. Thus deforestation, the negative impacts of mining, and land degradation, are difficult to address outside the main frame of climate change.

A third constraint on effective regional climate governance results from inter-agency rivalries. As noted above, in one sense each regional organization has a clear functional mandate but in practice there are elements of overlap, especially between the SPC and SPREP. This is compounded by the search for funds. A widely read report on climate financing in the region reported '[i]n an interview, one Pacific government official stated that there was 'a constant turf war between regional bureaucrats over who accessed the new pots of climate money' (Maclellan 2011: 37). The fate of the Pacific Islands Applied Geoscience Commission (SOPAC) illustrates the consequences of inter-agency rivalry. SOPAC was first established in 1972, formed initially as a UNDP Regional Project, then in 1990 as an independent intergovernmental organization. From January 2011, SOPAC was integrated as a new Division in the SPC. SOPAC's integration into the SPC was an outcome of the regional institutional framework reform process called for by the Pacific Island Leaders Forum. In 2007 the 38th PIF announced amendments to the regional institutional framework: 'the need to rationalise the functions of the Pacific Islands Applied Geoscience Commission (SOPAC) with the work programmes of the Secretariat of the Pacific Community (SPC) and the Secretariat of the Pacific Regional Environment Programme (SPREP) with the view to absorbing those functions of SOPAC into SPC and SPREP' (PIF 2007). This decision by PIF to rationalize SOPAC was in response to rivalries between SPC, SPREP, and SOPAC. Rivalries also exist between different PICTs, making regional cooperation difficult at times. Fissures are numerous and came up often in our interviews. These include different perceptions of urgency and capacity between atoll and non-atoll territories, distinctions between Francophone and Anglophone PICTs and their relationships with former and current colonial powers, the suspension of Fiji from the PIF between 2009 and 2014, and the shift to donors (New Zealand in particular) bypassing regional organizations in favour of bilateral aid on climate-related issues.

Conclusion

Climate change reinforces the notion of regional solidarity in the Pacific. The notion of a shared fate necessitates shared resources. The evolving architecture of regional governance is mobilizing resources and finance to combat climate change. On the surface it appears to be a successful manifestation of solidarity among states and territories facing a shared problem from a shared position of vulnerability. Solidarity has been institutionalized in three key regional organizations that mobilize resources for climate change action through distinct policy narratives about the place of the Pacific in global climate crisis. These organizations also distribute these resources to members. As a result climate finance in the Pacific is mostly generated by multilateral rather than bilateral donors. However, despite this apparent demonstration of regional agency, beneath the surface regional climate governance is heavily dependent on donors for finance and expertise. This exposes a dual vulnerability of PICTs. They are extremely vulnerable to the impacts of climate change while also being extremely vulnerable to growing dependencies on donors to address these impacts. We will discuss this further in Chap. 5.

Notes

1. http://www.paclii.org/cgi-bin/disp.pl/pits/en/treaty_database/1947/2.html?query=canberra agreement.
2. Until September 2016 membership was restricted to independent sovereign states but at the 47th Pacific Islands Forum Summit in Pohnpei it was unanimously agreed to grant full membership to French Polynesia and New Caledonia.
3. http://www.forumsec.org/pages.cfm/about-us/our-partners/crop/.
4. http://www.spc.int/en/about-spc/history.html.
5. In US dollars.
6. SPREP is also the secretariat for the Convention on Conservation of Nature in the South Pacific (Apia Convention), but the operation of the Apia Convention was suspended in 2006.

7. In US dollars.
8. http://www.sprep.org/Climate-Change/climate-change-overview.
9. http://gefonline.org/projectDetailsSQL.cfm?projID=2699.

References

Asian Development Bank (ADB). 2005. *Toward a New Pacific Regionalism*. Available at: http://www.adb.org/publications/toward-new-pacific-regionalism. Accessed 30 Aug 2013.

Baker, N. 2015. New Zealand and Australia in Pacific Regionalism. In *The New Pacific Diplomacy*, ed. G. Fry and S. Tarte, 137–148. Canberra: ANU E-Press.

Barbara, J., and H. McMahon. 2016. *Pacific Regional Youth Employment Scan*. Suva: Pacific Leadership Program.

Browne, C., and A. Mineshima. 2007. Remittances in the Pacific Region. IMF Working Paper (WP/07/35).

Chand, S. 2010. Shaping New Regionalism in the Pacific Islands: Back to the Future? (No. 61). ADB Working Paper Series on Regional Economic Integration.

Crocombe, R. 1976. *The Pacific Way: An Emerging Identity*. Suva: Lotu Pasifika Productions.

Federal Republic of Germany. 2016. *Climate Change Protection Through Forest Conservation in Pacific Island States*. Berlin: Federal Ministry for the Environment, Nature Conservation, Building and Nuclear Safety.

FFA. 2017. Economic and Development Indicators and Statistics: Tuna Fisheries of the West and Central Pacific Ocean 2016. Available at: https://www.ffa.int/system/files/FFA%20Economic%20and%20Development%20Indicators%20and%20Statistics%202016.pdf. Accessed 2 Aug 2017.

Frazer, I., and J. Bryant-Tokalau. 2006. Introduction: The Uncertain Future of Pacific Regionalism. In *Redefining the Pacific?: Regionalism Past, Present and Future*, ed. J. Bryant-Tokalau and I. Frazer, 1–24. Aldershot: Ashgate.

Harrison, D.H., and B.C. Prasad. 2013. The Contribution of Tourism to the Development of Fiji and Other Pacific Island Countries. In *Handbook of Tourism Economics, Analysis, Applications, Case Studies*, ed. C.A. Tisdell, 741–761. Singapore: World Scientific Publishing.

Herr, R., and A. Bergin. 2011. *Our Near Abroad: Australia and Pacific Islands Regionalism*. Barton: Australian Strategic Policy Institute.

Komai, M. 2013. *Reconfiguring Regionalism in the Pacific*. Port Vila: Pacific Institute of Public Policy.

Lawson, S. 2010. 'The Pacific Way' as Postcolonial Discourse: Towards a Reassessment. *The Journal of Pacific History* 45 (3): 297–314.

Maclellan, N. 2011. *Improving Access to Climate Financing for the Pacific Islands*. Sydney: Lowy Institute.

Makun, K.K. 2017. Imports, Remittances, Direct Foreign Investment and Economic Growth in Republic of Fiji Islands: An Empirical Analysis Using ARDL Approach? *Kasetsart Journal of Social Sciences* (Online First). https://doi.org/10.1016/j.kjss.2017.07.002.

Mara, R.K. 1997. *The Pacific Way: A Memoir*. Honolulu: University of Hawai'i Press.

Marawa, S. 2015. Negotiating the Melanesia Free Trade Area. In *The New Pacific Diplomacy*, ed. G. Fry and S. Tarte, 161–174. Canberra: ANU E-Press.

Mishra, S. 2005. Pacific Way. In *A Historical Companion to Postcolonial Though in English*, ed. Prem Poddar and David Johnson, 364–368. New York: Columbia University Press.

Pacific Community. 2016. *Pacific Community: Financial Statements for 2015*. Available at: http://www.spc.int/wp-content/uploads/2017/02/Annual-Report-EN-2015-V2.pdf. Accessed 31 Aug 2017.

Pacific Islands Forum (PIF). 2007. "Forum Leaders Communiqué" 38th Pacific Islands Forum, 16–17 October, Nuku'alofa, Tonga. Available at: http://www.forumsec.org/resources/uploads/attachments/documents/2007%20Forum%20Communique,%20Vava%27u,%20Tonga,%2016-17%20Oct.pdf. Accessed 21 Apr 2010.

Pacific Islands Forum (PIF). 2009. "Forum Communique" 40th Pacific Islands Forum, 5–6 August, Cairns, Australia. Available at: http://forum.forumsec.org/resources/uploads/attachments/documents/2009%20Forum%20Communique,%20Cairns,%20Australia%205-6%20Aug.pdf. Accessed 4 Dec 2011.

Pacific Islands Forum Secretariat (PIFS). 2003/2004. *Annual Report: Excelling Together for the People of the Pacific*. Available at: http://www.forumsec.org/resources/uploads/attachments/documents/2003-2004_PIFS_Annual_Report.pdf. Accessed 2 June 2011.

———. 2005a. *The Pacific Plan for Strengthening Regional Cooperation and Integration*. Suva: Fiji Pacific Islands Forum Secretariat.

———. 2005b. *The Pacific Plan*. Available at: http://www.sopac.org/sopac/docs/RIF/03_A_Pacific_Plan-2005.pdf. Accessed 22 Aug 2012.

———. 2010. *Pacific Plan Annual Progress Report*. Available at: http://www.forumsec.org/resources/uploads/attachments/documents/Pacific%20Plan%202010%20Annual%20Progress%20Report_Eng.pdf. Accessed 22 Sept 2013.

———. 2013. *Pacific Plan Review 2013: Report to the Leaders*. Available at: http://www.cid.org.nz/assets/Key-issues/Pacific-development/Pacific-Plan-Review-2013-Volume-2.pdf. Accessed 3 Aug 2014.

PIF. 1988. "Forum Communique" 19th Pacific Islands Forum, 20–21 September, Nuku'alofa, Tonga. Available at: http://forum.forumsec.org/resources/uploads/attachments/documents/1988%20Communique Tonga%2020-2%20Sept.pdf. Accessed 21 April 2010.

———. 1994. "Forum Communique" 25th Pacific Islands Forum, 31 July–2 August, Brisbane, Australia. Available at: http://forum.forumsec.org/

resources/uploads/attachments/documents/1994%20Communique Brisbane%2031Jul-2%20Aug.pdf. Accessed 18 June 2011.

PIFS. 2014. *The Framework for Pacific Regionalism*. Suva: Pacific Islands Forum Secretariat.

Rolfe, J. 2001. Peacekeeping the Pacific Way in Bougainville. *International Peacekeeping* 8 (4): 38–55.

Rosen, F., and P. Olsson. 2013. Institutional Entrepreneurs, Global Networks, and the Emergence of International Institutions for Ecosystem-Based Management: The Coral Triangle Initiative. *Marine Policy* 38: 195–204.

Shibuya, E. 2009. The Problems and Potential of the Pacific Islands Forum. In *The Asia-Pacific: A Region in Transition*, ed. J. Rolf, 102–115. Honolulu: Asia-Pacific Centre for Security Studies.

SPC. 2011. Euro 11.4 M Climate Resilience Project Will Help Nine Pacific Small Island States. 18 July, Secretariat of the Pacific Community. Available at: http://www.spc.int/en/component/content/article/740-114-m-climate-resilience project-will-help-nine-pacific-small-island-states.html. Accessed 7 Sep 2011.

SPC LRD. 2011. *SPC/GIZ Climate Protection Through Forest Conservation in the Pacific Islands*. 25 July, Secretariat of the Pacific Community Land Resources Division. Nabua: The Pacific Community Land Resources Division.

SPREP. 2014. *Secretariat of the Pacific Regional Environment Program Annual Report 2014*. Apia: Secretariat of the Pacific Regional Environment Program.

———. 2017. About PIGARREP. Available at: http://www.sprep.org/Pacific-Islands-Greenhouse-Gas-Abatement-through-Renewable-Energy-Project/about-piggarep. Accessed 31 Aug 2017.

Slade, S.G. 2009. *Statement by SG Slade at opening of 40th PIF*. Available at: http://lists.spc.int/pipermail/ppapd-fpocc/2009-August/000310.html. Accessed 25 Aug 2017.

Tarte, S. 2014. Regionalism and Changing Regional Order in the Pacific Islands. *Asia and the Pacific Policy Studies* 1 (2): 312–324.

Tavola, K. 2015. Towards a New Regional Diplomacy Architecture. In *The New Pacific Diplomacy*, ed. G. Fry and S. Tarte, 27–38. Canberra: ANU E-Press.

UNESCAP. 2013. *Economic and Social Survey of the Asia and the Pacific 2013*. Bangkok: United Nations.

CHAPTER 3

Constructing Climate Change in the Pacific

Abstract The prominence of climate change in the global arena has reconfigured the Pacific and its place in the world. Climate change has constructed the Pacific as the frontline of climate change. The first part of this chapter traces the emergence of climate change narratives and how they have been contextualized in the region and have established hegemony in political arenas. This is followed by an analysis of critical voices—many from within the region—that challenge climate hegemony and its impact, especially on other environmental and development problems faced in the region. The final section analyses the persistence of the frontline construction at the global level and the boost it offers to small island states in international climate politics.

Keywords Constructing climate change • Climate change hegemony • Climate change frontline • Pacific voices

Prior to his appearance at the 40th Pacific Islands Forum in Auckland in September 2011 the then UN Secretary General Ban Ki-Moon visited Kiribati. Among his activities was a visit to the village of Beki Ni Koora, where homes have been relocated because of rising sea levels. Here he stood on low-lying atoll and planted mangroves. While in Kiribati Moon stated, 'Many countries are dealing with climate change. But here, climate change threatens your territory, your culture and your very way of life.'

M. Williams, D. McDuie-Ra, *Combatting Climate Change in the Pacific*, https://doi.org/10.1007/978-3-319-69647-8_3

He also added, 'Kiribati will strengthen my belief, my conviction that this climate change is a much more serious security issue' (UNICEF EAPRO Media Centre 2011). The tree planting was a gesture towards adaptation, to stemming the tide. Yet the symbolism of Moon's visit is clear: the visit of the UN Secretary General—the figurehead for global cooperation—was a visit to the frontline of climate change.

Farbotko (2010) offers a different take on the ways PICTs are constructed in popular, diplomatic, and scholarly debates about climate change. She argues that atoll nations such as Tuvalu are experimental spaces for conceptualizing climate change at the global level; they are the 'canaries in the coalmine' (2010: 53). This is a powerful metaphor for the way many critical voices in PICTs perceive their position in climate change narratives. Farbotko adds,

> The metaphorical force of the canary in the coalmine rests with the idea that the canary – the Tuvalu islands – is not valuable in and of itself but rather is in service to a larger (global) environmental purpose. Even when the metaphorical death of the canary is considered to be lamentable, a rhetorically predictive manoeuvre is achieved whereby Tuvalu appears to be expendable. The disappearing islands thus embody not a located tragedy of importance in itself but a mere sign of the destiny of the planet as a whole. Tuvalu becomes a space where the fate of the planet is brought forward in time and miniaturised in space, reduced to a performance of rising seas and climate refugees played out for those with most control over the current and future uses of fossil fuels. (2010: 54)

The framing of PICTs as the 'frontline', as an experiment, and as a portent for the future has become a powerful narrative for imagining the region globally, internally, and in the international relations of PICTs. As Connell writes, they are 'sites where the great global narratives of climate change can be comprehended and made tangible and visible, attributed to distant sources, and confirmed by local "green governmentality discourses" based on indigenous knowledge, memory, and eyewitness accounts' (2015: 14). This is an enormous load to carry and one that, we argue, has eclipsed other issues affecting communities in the region. And while critical voices have emerged on the impacts of climate change narratives on everyday life, the visibility and agency of PICTs at the international level depended upon their adoption of this position, a position that enhances their presence and agency in arenas like the UNFCCC on the one hand and ensures the flow of aid and finance for climate-related adaptation on the other.

This chapter consists of three sections. The first one identifies the key narratives of climate change and how they come to be contextualized in the region. We argue that climate change narratives have framed the idea of the Pacific as the frontline of climate change and as a region bound by a shared fate. This orients the way the region is constructed at the global level. The second explores critical voices that challenge the dominance of these climate narratives. These are not the voices of climate deniers[1]; rather they are voices that challenge the simplicity of the 'canaries in the coalmine' view and the hegemony of climate change vis-à-vis other issues affecting communities and whole societies. The third section focuses on the ways in which PICTs utilize the dominant climate change narratives at the international level through a brief examination of their role in the Association of Small Island States (AOSIS). While the frontline view may be challenged within the region, in international arenas it is integral to the agency and moral power of PICTs, especially in climate change negotiations.

In short, this chapter interrogates competing policy narratives. It shows that the existence of a dominant narrative has not entirely eliminated the political space necessary for other narratives to emerge. Our argument is not that a contest of narratives is taking place but rather that in important ways the policy narratives are encoded in separate spatial domains. Policy narratives that emphasize vulnerability and extinction have the greatest traction in national and international arenas. Those that situate climate change within complex economic and social dynamics tend to reflect the realities of local communities and struggles emanating from local politics. It is equally important to note that we are not claiming that the 'frontline' narrative is somehow false or inaccurate. In drawing attention to the existence of critical narrations we are delineating a more complex political landscape than standard accounts.

Narratives of Climate Change in the Pacific

To understand how climate change in the Pacific is constructed and policy implemented by national governments and regional organizations we engaged in a critical analysis of various policy narratives developed by scholars, governments, and regional bureaucrats. The data corpus of this analysis is based on a review of the extant literature, government documents, and analyses by regional organizations. A critical reading of these texts illustrates the different ways in which climate change in the

Pacific has been conceptualized, and ways in which different narratives have influenced the ways that policy is conceived of and implemented on the ground. As discussed in the introduction, we define a policy narrative as a narrative or story about a specific policy problem designed to influence decision-making and shape policy implementation. These narratives create shared meanings for decision-makers both within and outside of a polity who are concerned with the same problem set. Thus in the case of climate change in the Pacific these policy narratives link decisions at the local, national, regional, and transnational levels.

Our focus on the evolution of key narratives and the resultant policy response is not meant to deny the urgency of responding to the threats posed by climate change for the PICTs. Indeed, we subscribe to the various analyses that document the severity that climate change poses to Pacific Island societies. Nevertheless, a critical exploration of the role of regional cooperation in responding to the threats posed by climate change must begin not from a statement of the problem to be analysed but through an analysis of the 'origins' of the problem. In other words, key questions such as when was climate change recognized as a problem/ policy issue for PICTs and how was the problem defined and possible solutions articulated cannot be ignored.

Sea Level Rise

The dominant narrative on the implications of climate change and its effects on the PICTs was at first largely, and still is to an extent, focused on sea level rise and the loss of land and/or of whole Pacific Island nations. As noted by Connell (2003: 91), climate change first became an international concern in the 1980s as the global scientific community began to take note of global warming and sea level rises that would affect low-lying islands. By about the 1990s there were a number of academic literatures on the effects of sea level rising on atoll nations such as Pernetta and Hughes (1990) and Roy and Connell (1991). A recent report projects that by 2100 sea level rise in the Pacific will likely vary between 0.26 to 0.55 metres for low emissions and 0.52 to 0.98 metres for high emissions (IPCC 2013). And another recent study has shown how the El Niño effect will negatively impact sea level rise, flooding and tropical cyclones (FRDP 2016). Sea level rise is important as it locates PICTs on the frontline in the global climate change narrative, especially in the last decade. Indeed as Connell has argued, '[m]assive public discussion of climate change at international, regional, and local levels has contributed to (Sea Level Rise) being the primary explanation for harmful, unusual, or unprecedented environmental changes' (2015: 2).

Most sources point to the Intergovernmental Panel on Climate Change's (IPCC) first report in 1990, and the development and formation of the AOSIS at the Second Climate Conference as a turning point for global recognition of the challenges facing PICTs (see Shibuya 1997; Tisdell 2008; Kendall 2012). The IPCC report underscored the dangers of climate change and the possibility of atoll nations being flooded by 2030. Shibuya (1997) and Betzold (2010) also argue that statements by key actors brought further attention to the plight of the PICTs, such as Robert Van Lierop, Vanuatu's UN ambassador, and strong statements such as 'we don't have time to wait for conclusive proof. The proof, we fear, may kill us' (Betzold 2010: 138).

The notion of a shared fate drives the narrative for regional cooperation. As early as 1983 a UNEP report noted that the imperative of regional cooperation on 'environmental matters' arose from a shared heritage (Dahl and Baumgart 1983: 25). As the contributors to a later UNEP report noted sea level rise affects the coastal ecosystems of all Pacific Island nations (see Pernetta and Hughes 1990). For example, this shared vulnerability provides the rationale for the creation of the Regional Seas Action Plans. The centrality of a shared fate is represented by Grasso (2006) who suggests constructing a regional Pacific agreement to work in conjunction with the Kyoto Protocol, which creates a distinction between mitigation and adaptation strategies.

Possible Extinction

A second key policy narrative focuses on the possible extinction of Pacific Island nations and its peoples. A number of studies have raised the extinction scenario. For example Edwards (1996: 69–70) contends that climate policy for the Pacific is compromised because these islands will not exist in the future. As Nunn laments, sea level rise 'will spell the end of today's Pacific in many ways: a radical change in geography spawning fundamental changes in settlement patterns and communications infrastructure, subsistence systems, societies and economic development' (2013: 165). At the local level the narrative emphasizes the threat that climate change poses to traditional cultures and socio-economic structures. At the national level it contends that sovereignty will be compromised. For example it has been argued that since atoll states are largely ethnically 'homogenous' with 'high population density … there is little political distance between the people and the nation-state' (Barnett and Adger 2003: 327; see also Corbett 2015). A threat to the livelihoods and human rights of the inhabitants of atoll countries is therefore a threat to the 'nation-state' itself (Barnett and Adger 2003: 327).

Migration

Migration features widely in discussions of climate change and the Pacific and is at the centre of a third policy narrative. Here the focus is dual. On one hand, some contend that migration poses a threat to the existence of communities. It has been argued that climate change threats can lead to migration, either to already saturated urban areas, or to neighbouring islands, potentially leading to poverty—due to limited labour markets—and a potential loss of culture (Edwards 1999: 316–17). On the other hand, others are more concerned with the possible threat that migration may pose to relations between the PICTs and Australia and New Zealand. Moore and Smith (1995: 119) argue that migration flows will depend on the type and immediacy of the impacts caused by climate change, on neighbouring countries' willingness to welcome climate refugees, and on the respective states' 'adaptive strategies'. Wyeth (2014) argues that in the case of Kiribati, bilateral migration arrangements with Australia and New Zealand is the best policy, as they are the wealthiest in the region, have low-density populations and they already have a long history of migration. Strokirch (2007: 555) also lists Pacific Islanders from countries such as Tokelau and Tuvalu as inevitably having to migrate to New Zealand, 'the Marshall islanders who will go to the United States under the terms of the Compact of Free Association and Kiribati, does not have anywhere as of yet to go.' Politicians also engage in these conversations. This is illustrated through Barnett (2003: 12) who quotes the comments of Prime Minister of Tuvalu to the *Fiji Times* in February 2000 that 'Tuvaluans are seeking a place they can permanently migrate to should the high tides eventually make our homes uninhabitable'. We will discuss this further in the following section.

Livelihoods

A fourth narrative develops the theme of livelihoods. This narrative links the damage to ecosystems with agricultural production and food security. Fisher (2011) uses the example of Tuvalu to argue that 'biophysical climate changes combine with human systems to generate a series of interconnected and cascading effects that affect social vulnerability'. That is, the effects of climate change merge with and affect social security systems, such as 'sanitation, development, health, food security, quality environment, political capacity, resources, and ecosystem biodiversity' (Fisher 2011: 311). In linking climate change with food security Barnett argues that climate change in the Pacific is indeed dangerous on an economic level, given its likely effect on agricultural production and fisheries, but also on subsistence of the population. He asserts that 'through its impacts on production, the ability of countries to import food and the ability of households to purchase food, and human health, climate change puts at risk the very basic and universal need for people in the islands to have access to sufficient, safe, and nutritious food at all times' (Barnett 2005: 236).

Gender

Climate change has a complex relationship to gender. Narratives of climate change and gender remain marginal, but speak to three main dynamics. First, the impacts of climate change are gendered in that they affect women and men in different as well as similar ways. Second, the ways climate change is governed in the Pacific reflects existing gender relations in local and national decision-making as well as the status of particular expertise over others. Third, participation in political arenas at the national and regional levels remains heavily gendered. Lane and McNaught (2009) points to a new type of literature that focuses on gendered approaches to climate change that was first identified internationally through the 52nd Commission on the Status of Women during the Interactive Expert Panel in 2008 that has until now largely been ignored. Lane and McNaught (2009) through an empirical analysis stipulates that a gendered approach to climate change adaptation is required as there are gender-specific impacts and knowledge of climate change. Hayward (2008: 89) also briefly refers to the importance of women in the PICTs and their vulnerability to climate change and the lack of literature on this issue. Overall, however, presently there is very little literature on this issue and it is often only mentioned briefly.

Security Threat

The literature on climate change as a global security threat is a fairly new phenomenon. Kendall (2012) argues that this is because traditionally security has been seen through a realist paradigm, whereby security is interpreted as an extraterritorial military threat, which did not look at issues of environmental and human security. Dupont and Pearman (2006: viii) also concur with this analysis and argue that environmental security has largely been ignored and underestimated in public policy, academia, and the media; however, as inter-state wars have diminished, non-state threats such as terrorists, infectious diseases, and unregulated migration have entered the discourse of international security. Barnett (2003) also traces the evolution of the environmental security narrative and argues that it started in 1971 when Falk's book *This Endangered Planet* was published linking the environment to the politics of security. Barnett (2003, 2005), however, suggests that this narrative did not reach the international political scene until the Toronto Conference in 1988 labelled 'The Changing Atmosphere Implications for Global Security', which specifically identified climate change as an 'environmental security' problem. Since then actors from PICT leaders, to the UN Secretary General, to the defence departments in countries like Australia have made the connection and in these connections the Pacific is noted for its fragility and potential for instability. A central theme of this book is a focus on the security dimension of climate change and we will therefore discuss this at length in the following chapter.

Critical Voices

While these hegemonic policy narratives have not encountered major challenges this does not mean that dissenting voices and analyses that differ from the mainstream are not to be found in the literature. Here we focus on literature that challenges the way climate change has been narrated in the Pacific. We are not referring to climate change denial, but rather to voices that deconstruct the narratives of climate change and the way they have shaped politics in the region. Whilst the impacts of climate change have been highlighted and accentuated, especially in relation to sea level rises, only a small portion of the literature is on solutions for the PICTs. Of the literature from the academic and media circles that do focus on enacting policies in the PICTs to respond to the problems presented by climate change, they have deliberately concentrated on adaptation strategies in the PICTs. For example, Bell (2013) argues that climate change is mostly as a result of developed industrialized countries' emissions of greenhouse gases and it is therefore the responsibility of these polluters to develop mitigation policies to curb anthropogenic climate change.

Moreover, Strokirch (2007) points to the PICTs having to formulate adaptation strategies regardless if industrialized nations adopt mitigation measures, as there are some effects of climate change that are inevitable.

Kempf (2009: 200–201) contests the hegemonic discourses that represent the Pacific as helpless, and argues that while migration due to climate change may be the ultimate solution for atoll states, adaptive strategies should not be undermined and voices from the Pacific should be heard. Concerning migration as a climate change adaptive strategy, Mortreux and Barnett (2009: 105) argue that the absence of sufficient focus on 'the capacity of social and ecological systems to adapt, the constraints and barriers to adaptation, and the costs of and limits to adaptation' means that conclusions on climate refugees are incomplete. In their view, the widely held notion that the low-lying atoll islands are doomed need not hold true, if measures are taken to reduce greenhouse gas emissions, and adequate adaptation strategies are developed (Mortreux and Barnett 2009: 106). Further, they warn that dominant discourses of migration as a result of climate change can pose a future danger, in that 'large-scale migration may be an impact of climate change affected by policy responses in anticipation of climate impacts rather than by material changes in the environment *per se*' (Mortreux and Barnett 2009: 111). Moreover, Barnett has challenged the dominant assertion that all PICTs will suffer the same fate. He argues that the effects of climate change are complex, and it cannot be assumed that all island states will become 'Titanic states'. A focus on solutions, such as a reduction of greenhouse gas emissions and sustainable development is essential, rather than deeming the Pacific a doomed region (Barnett 2005: 216–17).

Much attention has been given to the possible extinction of PICTs. Most of the studies are on atoll countries such as Tuvalu and Kiribati, which are most likely to be the first to be inundated, as a consequence of climate change. At the same time however, there is a growing academic literature that questions depictions of PICTs as weak and dependent on other states to counter climate change threats. Connell (2003), by focusing on Tuvalu, highlights the alarmism played out by NGOs and the media that use atoll nations such as Tuvalu in their awareness raising. He points to the Tuvaluan government playing the same vulnerability card as the media. Farbotko (2005) adds to the literature by critiquing the 'vulnerability' discourse through climate change sensationalism adopted by the Australian media. Farbotko (2005) through a discourse analysis of the articles published by the *Sydney Morning Herald* on Tuvalu finds that

Tuvaluans are portrayed as helpless 'tragic victims' that will become environmental refugees as a result of global warming. Farbotko (2005) argues that this type of journalistic narrative that is based on an imminent tragedy downplays and impedes on alternative discourses that allow for resilience strategies from being implemented. Barnett (2001) also acknowledges the problem of this vulnerability discourse that stereotypes these states as weak and argues that a strategy recognizing resilience is required.

Thus, whilst the literature in the media and, to an extent, the academic journals are based on the PICTs being vulnerable, weak, and imminently about to disappear, there is literature that questions this dominant policy narrative and demonstrate that there are political dimensions at play, whereby politicians from the PICTs take advantage of this narrative of fragility to gain support from industrialized nations. This will be discussed in the following section.

Perhaps the most critical of all interventions comes around migration, particularly literature that challenges the assumption that migration is the only 'adaptive' solution to climate change in the Pacific. For example, Mortreux and Barnett (2009) argue that climate change will not inevitably lead to whole national populations migrating. Moreover, they find in their research that out of the Tuvaluans that they interviewed, they themselves did not list climate change as a reason to leave the island. Other accounts that seek understandings of climate change from within Pacific communities 'on the ground' rather than those representing PICTs at the regional and transnational levels reveal a range of positions on climate change that challenge dominant narratives.

In his research on the Marshall Islands, Rudiak-Gould argues that climate change narratives give a name to experiences that have shaped island communities for generations. He suggests that observations about local change are now articulated in the language of climate change as a result of the 'reception of scientific narratives' (2016: 265). He discusses the work of a local NGO, the Marshall Islands Conservation Society, which markets itself locally as a cultural organization and as an environmental organization to outside donors who, 'understand the language of "biodiversity" and "conservation" better than the language of *ko- jparok manit* ("maintaining tradition") and *lale doon* ("taking care of one another")' (2016: 265). He adds, 'in the Marshall Islands, climate change is not so much an environmental threat as it is an existential one' (2016: 265).

For other communities it is also an economic one. Most countries in the region are not only poor but their economies are dependent on a limited number of sectors for foreign exchange and investment. PICTs are often described as having 'MIRAB' economies. Initially developed by Bertram and Watters, MIRAB refers to a reliance on migration, remittances, aid dependency and bureaucracy, which has both dominated island economies and 'determined their evolution' from colonial export oriented economies since the 1950s (1986: 47). MIRAB challenged the notion of 'vulnerability'—not in the sense now widely understood through the climate change paradigm, but rather as exposure to economic shocks being experienced elsewhere in the post-colonial world. MIRAB economies, they argued, reveal responses to opportunities and changing regional and global politics. At the time the concept challenged the growing orthodoxy of dependency theory, and highlighted the agency of PICTs to insulate themselves from external shocks. Bertram has developed the concept further in intervening years, commenting in 1999 that there are variations in the mix of these factors in different parts of the region and that they perhaps idealized the role of kin networks (1999: 125), and in 2006 that while empirically most states they focussed on still fit the MIRAB model, remittances and aid flows are more vulnerable and erratic than first imagined (2006: 4). However, the model has provided fuel for debate about PICTs, their economic vulnerabilities, and—more recently—the extent to which climate change completely recalibrates the notion of vulnerability in the region (Barnett 2001; Connell 2010; Kelman et al. 2015). However, as Lazrus argues, many aspects of MIRAB economies indicate that PICT economies, social networks, and resources stretch far beyond the physical boundaries of the region, drawing our attention to the 'totality of relationships that constrain and support the agency of islanders to navigate climate change' (2012: 295). In other words, it is not simply climate change that induces migration and, given the networks and larger geographic worlds in which Pacific Islanders are embedded, Pacific lives are not and have not been solely contained within the region.

Climate change, however, continues to shape the ways the region is perceived and the ways development proprieties are determined. In a relatively early article on the impacts of climate change on development in the pacific, Connell writes (1993: 188),

> If adaptation [to climate change] is the most rational political option, the climate problem tends to fade ('Chalk on the White Wall') compared to the already extremely serious and immediate social and economic problems

> confronting developing countries. As indicated, such problems including rising populations, environmental problems including sewage and garbage disposal, pollution and over-use of underground water supplies, the consumption of energy and pressures on housing space. Infrastructure is already stretched to its limit. Thus climate-oriented policies to cope with climatic change tend to become part of development policies in general, and are unlikely to be prominent amongst them.

As Barnett and Campbell note almost two decades later, in the Pacific 'climate change has developed its own cadre of experts … [who] circulate in a science-policy bubble that at times floats far above where impacts will be felt and adaptions required' (2010: 179). They add that as a result of this cadre, 'climate change tends to dominate *everything* so that it seems like it is the paramount environmental problem, vulnerability is the most pressing social problem, [and] adaption should be the goal of development and resource management policies' (2010: 179–80). Environmental issues that challenge core aspects of PICT economies are even harder to politicize.

During fieldwork respondents would often raise this problem. During a group interview with respondents from the UNDP in 2009, particular frustration was expressed with the lack of funding and will to address issues like urbanization and waste management owing to their seeming lack of importance to the agenda set by climate donors and the community of experts and consultants embedded in the region (Interviews, Suva, June 2009). For many communities and activists in the region, the impacts of rapid urbanization—crowding, waste, pollution, crime, poverty, disease, cost of living, and access to services—are exemplars of what is lost in the focus on climate change. For decades concerns over urbanization in PICTs has risen alongside concern over climate change (see Bryant-Tokalau 1995; Connell 2011; Connell and Lea 2002; Thornton et al. 2010). And while the two are connected, as certain local ecologies in PICTs become more difficult to inhabit migration to urban areas within and between PICTs grows, there are multiple factors driving urbanization in PICTs, many of which are common to other contexts. As Connell argues, urbanization in PICTs has become an 'elephant in the room', which is ignored in policy and practice, perhaps an aberration, a circumstance in transit—hopefully not really there and surely not the 'real Pacific' (2011: 121).

Urbanization is driven by what Storey and Hunter (2010) call 'a perfect storm' of factors. Their study of urbanization in Tarawa, Kiribati, examines factors ranging from reactions to environmental degradation such as water shortage—exacerbated by climate change—to the harmful effects of development projects. They close by concluding 'that there are real threats to SIDS posed by predicted climate change and sea level rise is not disputed, but even in small island states the "brown agenda" issues of pollution, sewerage and solid waste management cannot afford to be left outside of the concerns of governments, development agencies and donors' (2010: 177). This plea is now commonplace by scholars working in and hailing from the region. The hegemony of climate change vis-à-vis other development and environmental issues has also been discussed in some of the more critical studies in various PICTs. These authors do not deny the severity of climate change, a necessary and now reflexive caveat, but are simultaneously critical of the impact the climate change discourse and attendant policy fixes is having on the ways PICTs are governed, financed, and imagined in a world of shared fate.

Using the case of Tuvalu, McCubbin et al. (2015) locate climate change in a complex set of forces impacting on social life for individuals and communities. They argue that climate change-related forces interact with non-climate change-related forces (or those with only a very indirect link) as 'multiple stressors'. They note that among their respondents, 'While climatic conditions were also identified as influential, these (with the exception of water) were noted less frequently than the economy, food, overcrowding and culture. This finding is especially noteworthy given the widespread assertion that climate change and sea-level rise are the main drivers of vulnerability in Pacific Islands, particularly in atoll countries like Tuvalu' (2014: 47). However, climate was seen as exacerbating stresses around water, land, and food, themselves transforming as urbanization and changes to land tenure take hold (2014: 49–52). For their respondents, climate change is bound up in other stresses and there is limited logic in planning adaptation strategies that deal only with climate-related impacts.

Kelman (2014) argues that climate change can 'depoliticise' the development challenges faced by small island states. In other words, development challenges are no longer subject to debate, action, and contention; rather they are bound up with climate change in different ways. Kelman focuses on three 'themes' of depoliticization. The first is the emphasis on climate-related hazards rather than all hazards affecting island states. Not only do other development challenges become related or obscured internally and to donors, the hazards related to climate change are presented as new and thus the experience of communities and polities in dealing with past hazards are bypassed in favour of (usually) imported expertise (2014: 123). The second is that long-term development challenges are less pertinent as the 'frontline' view suggests there will be no long-term in many PICTs. This is especially pronounced in writing on migration, which has multiple causal factors in PICTs including—but not limited to—climate change (2014: 124). Yet discussion of migration tends to focus on climate refugees or climate-induced migration and not the other factors driving migration in the region: limited livelihoods, poverty, land grabs and degradation, and the culture of remittance. The third is missed opportunities to reduce vulnerabilities to all kinds of shocks, not just those framed through climate change (2014: 126).

The focus on climate change presents a limited view of the economic, political, and environmental relationships PICTs have with each other and with other states, firms, and investors. There is also limited consideration of formal and informal authority, particularly customary law, which can be crucial in mobilizing community support for climate-related adaptation policy and addressing the impacts of climate change and other factors on local ecologies (Dyer 2017). For instance, in his analysis of forests in the Solomon Islands (Marovo), Hviding explores the relationships between representatives of the 'West', 'mainly conservation-focused NGOs and ecology-oriented tourists', the 'East', 'mainly Asian agents of transnational capitalism', and the 'Rest', local Marovo communities and their entanglements with customary authority and government (2003: 593). Hviding posits 'in the South Pacific of today, the East engages intensely with the Rest, while the West has taken on a definitely diminished role compared with the state of affairs in colonial times' (2003: 593). Hviding's exploration of forest politics in Marovo—while not dealing directly with climate change—suggests that all environment discourses are interpreted locally, and while agents of the West (and in climate change aid and finance Japan and South Korea are important inclusions in this category) promote environmental norms and climate adaptation policies and projects in the Pacific, agents from elsewhere—in this case logging and fishing firms from China—engage in extractive and potentially hazardous relationships with the same communities and governments. This is a challenge to the notion that PICTs accept dependency.

Other studies have identified similar dynamics in other sectors, such as fishing, mining, logging, and oil and natural gas exploration; a situation made even more contentious as the Pacific is often a battleground between states pursuing other agendas, such as between China and Taiwan over influence and external recognition (Atkinson 2010; Brant 2013; Shie 2007; Yang 2009). It is crucial to note that a geographic polarity—'western environmentalism versus eastern extraction'—does not always hold and there are certainly state and non-state entities from 'the West' engaging in environmental degradation in PICTs while bilateral and multilateral aid seeks a reduction in environmental harms and accelerated climate change adaptation. An exploration of the full extent of environmental issues is beyond the scope of the point being made here, rather we seek to point out that while PICTs prepare for climate change adaptation and mitigation funding from particular donors in multilateral and bilateral relationships, they also engage in environmentally destructive resource extraction emanating from other bilateral relationships.

Indeed, climate change requires of PICTs what Grydehøj and Kelman (2017) refer to more generally in small island polities as 'conspicuous sustainability', namely the 'initiatives undertaken in the name of sustainability and climate change mitigation that also seek to gain competitive advantage, strengthen sustainable tourism or ecotourism, claim undue credit, distract from failures of governance, or obviate the need for more comprehensive policy action' (2017: 107). It is precisely because PICTs are on the 'frontline' of climate change, cast as remote and vulnerable, that conspicuous sustainability is so beneficial. It is also competitive and in the drive to demonstrate sustainability many islands risk being trapped in 'struggling to meet potentially unrealizable expectations they have foisted on themselves' (2017: 111).

As climate change comes to almost wholly define the Pacific as a region to certain audiences, PICTs as polities, and the future/extinction of Pacific societies other pressing issues are either ignored or absorbed into the ways climate change is addressed, governed, and financed. Scale matters here. In relatively small polities the hegemony of the climate change agenda pushes other issues out of smaller political and economic arenas. However, alternative discourses are emerging from the Pacific, discourses that do not contest the realities of climate change but that challenge the homogeneity of the 'frontline' imaginary. The canaries in the coalmine, to return to Farbotko's metaphor, are singing various tunes, and are very much alive. What is clear is there is some discord, some dissonance between the experiences and fears of communities and the official position of governments and regional organizations in PICTs; positions shaped by the regional and global politics of climate change.

The Global Appeal of Vulnerability

If we accept that narratives of climate change have been highly influential in PICTs, constructing an imagination of states, territories, and communities on the 'frontline' of climate change and as the proverbial canaries in the global coalmine, and if we acknowledge that there are alternative narratives about climate change that do not necessarily reject its severity but seek to locate it in existing political and social contexts in PICTs, what are we to make of the positions taken by PICTs in international arenas for discussing climate change? We argue that while complex and nuanced narratives around climate change are emerging in the region itself, the dominant narrative of vulnerability and extinction continues to be presented by PICTs in international arenas. The strategy appears to have two components. First, to emphasize the severity of the issue in the region and garner resources to combat climate change and second, to build and reinforce solidarity among PICTs and between PICTs and other island states.

In this section we explore briefly the role of the AOSIS as a regional framework for Pacific nations. Conceived in 1990 as a negotiating cluster in order to address the adverse effects of climate change under UNFCCC (AOSIS n.d.; Betzold et al. 2012: 591) AOSIS has provided the PICTs with both extra-regional allies and contributed to the emergence and development of a specific regional identity. AOSIS is largely driven by the fact that, although its 44 members equate a mere 5% of the Earth's population and contributed the least towards the causes of climate change, they are nonetheless the most vulnerable (Ashe et al. 1999; Brindis 2007; Davis 1996). Climate change affects different members of the coalition differently. For instance, low-lying atoll states such as Kiribati or Tuvalu face threats of unsustainability to the point of destruction, whilst other islands such as Cuba or Belize could adapt (Betzold et al. 2012: 594). Nonetheless, all member states have common geographical characteristics (low-lying island states) and face similar development challenges, which render them disproportionately susceptible to environmental changes (AOSIS n.d.; Betzold et al. 2012: 594; Ashe et al. 1999: 210). Since it is not recognized as an independent group, AOSIS is also a member of the G77 and operates without a charter (Betzold 2010: 135).

Since its conception, AOSIS has played a major role in UN climate change negotiations. Part of AOSIS' power to direct attention towards climate change negotiations was its stand as victims of developed countries' actions (Betzold 2010: 138). The moral weight of AOSIS' plight landed the Alliance a seat on the 'Bureau of the negotiations', in the Intergovernmental Negotiating Committee as well as in the Conferences of the Parties (COP) (Betzold 2010: 138). In 1994, the Global Conference on the Sustainable Development of Small Island Developing States was held, resulting in the adoption of the Barbados Programme of Action for the Sustainable Development of Small Island Developing States (UNGA 1994). This programme expanded from the initial climate change demands to cover sustainable development (Chasek 2005: 137). Reaffirming the Barbados Programme in 2005, after the effects of the 2004 tsunami, AOSIS drafted the Mauritius Declaration, which 'outlines the SIS struggle to exist in the face of the threat of climate change' (Brindis 2007: 45).

In 1995 AOSIS presented a protocol requesting that Northern governments reduce CO_2 emissions by 20% and lower greenhouse gas emissions by the year 2005 (IISD 1994) as well as develop 'a coordination mechanism for cooperation on economic, administrative and other implementation measures' (IISD 1994). The outcome, the Berlin Mandate, states that developed countries must 'elaborate policies and measures', and 'set quantified limitation and reduction objectives within specified time-frames, such as 2005, 2010 and 2020, for their anthropogenic emissions by sources and removals by sinks of greenhouse gases not controlled by the Montreal Protocol' (UNFCCC 1995: 5).

PICTs have been key players in the conception of AOSIS, and in the continuation of diplomatic focus on the effects of climate change on Small Island Developing States (SIDS) (Taplin 1994; Davis 1996). In terms of impact, the case of the Pacific region is particularly strong, because of the number of low-lying atoll island states in the region whose very existence is threatened by climate change. At the leadership level, the Pacific has made important contributions to AOSIS. For example, four out of the total of nine AOSIS chairpersons have come from Pacific Islands. The very first chair of AOSIS in 1990 was the UN ambassador from Vanuatu, Robert Van Lierop, who played an important role in setting the 'stabilisation of greenhouse gas emission' in the UNFCCC (Betzold 2010: 137). Arguably, van Lierop's leadership ensured 'substantial consideration to the concerns of South Pacific and small island states' (Taplin 1994: 279–80). In the first 1995 COP, it was Samoa that spoke for AOSIS, and called for the AOSIS protocol (later the Berlin Mandate) to be adopted (IISD 1994). In 1989, at the IPCC's Working Group III, Vanuatu's representative Ernest Bani addressed developed nations urging them to 'formulate and implement more judicious environmental policies in order to prevent us [the South Pacific islanders] from becoming endangered species or the dinosaurs of the next century' (in Taplin 1994: 274). A year later, at the Second World Climate Conference in Geneva, following warning signs of the consequences of sea level rise, Tuvalu's Prime Minister, Bikenibeu Paeniu, made an important statement regarding the responsibility of industrialized countries to SIDS (Taplin 1994: 274; Davis 1996). Importantly, Paeniu included in the statement not only the Pacific SIDS, but also the Caribbean and the Indian Ocean islands, thus creating a link between the regions based on their shared vulnerability (Taplin 1994: 274). Similar statements have been made at each of the various COP meetings, though as Rose notes, PICTs priorities 'can be seen to have shifted towards securing financial resources to enable them to take action to effectively adapt to the impacts of climate change that have already commenced and are predicted to become acutely serve in the future' (2015: 213). For instance, the closing statement by AOSIS at the COP 21 in Paris in 2015 spoke of the threat of a 3° temperature rise to the very existence of many SIDS, the time pressure to act, and the need for resources. From the statement by Chair Ahmed Sareer of the Maldives

> ...tackling climate change and adapting to its impacts will require significantly scaled-up, new, additional and predictable financial resources, starting from the minimum of $100 billion USD per year by 2020. This should include special provisions to enhance access by SIDS, especially to public, grant-based support for adaptation, given the particular challenges and the attendant existential threat that climate change poses to our countries. For AOSIS, finance is critical to effective implementation, and in light of our capacity constraints, simplified access is essential. (AOSIS 2015: n.p.)

For PICTs, AOSIS offers a way to participate in the key arenas of climate change politics in solidarity with other states and territories and away from the power of regional hegemons and donors such as Australia and New Zealand. Perhaps most remarkable for PICTs was the role of Tuvalu at the COP 15 in Copenhagen in 2009. After the G77 split over whether to pursue a new agreement or to continue with a revised Kyoto Protocol, Tuvalu pushed for a new agreement that would demand deeper global emissions cuts by developed and developing countries. While the proposal was rejected, the negotiations were suspended and Tuvalu was lionized among NGOs and other climate activists at the conference. As Benwell put it, 'their emotional rhetoric and capture of formal rules of negotiation procedure combined in these circumstances to make them powerful independent players in the climate change regime' (2011: 205).

Tuvalu's leverage at the COP was—in a sense—remarkable given a country of 14,000 people was able to halt international negotiations and lead opposition to powerful actors in the G77 plus China alliance. On the other hand, the agency of Tuvalu at the global level is tied specifically to the dominant policy narratives shaping this issue. The arena and the issue within which countries like Tuvalu can set or at least challenge the agenda of other states arises precisely because they are on the frontline, because they are the canaries in the coalmine, and because they have a platform through solidarity with other states and territories sharing a similar fate. In their analysis of AOSIS negotiations at COP 15, Deitelhoff and Wallbott write, 'It is unlikely that a small island like Tuvalu would have been able to cause a stalemate in the negotiations if it were not for the support of its coalition members ... [and] AOSIS's coherence and institutional persistence in the face of major powers' pressure represent a success' (2012: 361). Therefore, despite critical voices in the Pacific, the value of the frontline construction and the narratives that underpin persists for governments and their alliances of solidarity

Conclusion

Narratives of climate change have been internalized in the Pacific. These narratives mesh climate science with observations of change in the region with policy prescriptions from donors, and in the international climate change arenas more generally. These narratives have produced and continue to reproduce the Pacific as the frontline of climate change, and PICTs as the canaries in the coalmine. This has become hegemonic making it very difficult to frame other issues and vulnerabilities internally and externally to the region. There are, however, critical voices. These voices do not deny the realities of climate change but identify the complexities of contemporary life in the Pacific, complexities overshadowed—and perhaps depoliticized—by the focus on climate change. For developing countries with heavy dependency on aid and remittances undergoing rapid urbanization and ruptures to agrarian and other livelihoods, more conventional development problems are being relegated to the margins of political discourse, while the focus on the very existence of PICTs—especially atoll countries—takes centre stage. While critical voices demonstrate a contentious politics of climate change at the local level, at the international level PICTs remain invested—and to some extent dependent—on the notion of extinction for leverage and traction in global climate negotiations, responding in part to the political opportunities presented and in part to the grim realities of sea level rise, unpredictable weather, and climate insecurity.

Notes

1. Such as Bjorn Lomborg (2016) and his widely distributed paper 'About Those Non-Disappearing Pacific Islands'.

References

AOSIS. n.d. About AOSIS. Available at: http://aosis.org/about-aosis/. Accessed 19 Feb 2014.

AOSIS. 2015. *AOSIS Opening Statement for 21st Conference of Parties to the UNFCCC*. Available at: http://aosis.org/wp-content/uploads/2015/12/FINAL-AOSIS-COP-Statement-Paris-.pdf. Accessed 28 July 2016.

Ashe, J.W., R. Van Lierop, and A. Cherian. 1999. The Role of the Alliance of Small Island States (AOSIS) in the Negotiation of the United Nations Framework Convention on Climate Change (UNFCCC). *Natural Resources Forum* 23: 209–220.

Atkinson, J. 2010. China–Taiwan Diplomatic Competition and the Pacific Islands. *The Pacific Review 23* (4): 407–427.

Barnett, J. 2001. Adapting to Climate Change in Pacific Island Countries: The Problem of Uncertainty. *World Development 29* (6): 977–993.

———. 2003. Security and Climate Change. *Global Environmental Change 13* (1): 7–17.

———. 2005. Titanic States? Impacts and Responses to Climate Change in the Pacific Islands. *Journal of International Affairs* 59 (1): 203–219.

Barnett, J., and W.N. Adger. 2003. Climate Dangers and Atoll Countries. *Climatic Change* 61 (3): 321–337.

Barnett, J., and J. Campbell. 2010. *Climate Change and Small Island States: Power, Knowledge, and the South Pacific*. London/Washington, DC: Earthscan.

Bell, D. 2013. Climate Change and Human Rights. *Wiley Interdisciplinary Reviews: Climate Change* 4 (3): 159–170.

Benwell, R. 2011. The Canaries in the Coalmine: Small States as Climate Change Champions. *The Round Table* 100 (413): 199–211.

Bertram, G. 1999. The MIRAB Model Twelve Years On. *The Contemporary Pacific* 11 (1): 105–138.

———. 2006. Introduction: The MIRAB Model in the Twenty-First Century. *Asia Pacific Viewpoint* 47 (1): 1–13.

Bertram, G., and R.F. Watters. 1986. The MIRAB Process: Earlier Analyses in Context. *Pacific Viewpoint* 27 (1): 47–59.

Betzold, C. 2010. 'Borrowing' Power to Influence International Negotiations: AOSIS in the Climate Change Regime, 1990–1997. *Politics* 30 (3): 131–148.

Betzold, C., P. Castro, F. Weiler. 2012. AOSIS in the UNFCCC Negotiations: From Unity to Fragmentation?. *Climate Policy* 12 (5): 591–613.

Brant, P. 2013. Chinese Aid in the South Pacific: Linked to Resources? *Asian Studies Review* 37 (2): 158–177.

Brindis, D. 2007. What Next for the Alliance of Small Island States in the Climate Change Arena? *Sustainable Development Law and Policy* 7 (2): 45–85.

Bryant-Tokalau, J.J. 1995. The Myth Exploded: Urban Poverty in the Pacific. *Environment and Urbanization* 7 (2): 109–130.

Chasek, P.S. 2005. Margins of Power: Coalition Building and Coalition Maintenance of the South Pacific Island States and the Alliance of Small Island States. *Review of European, Comparative and International Environmental Law* 14 (2): 125–137.

Connell, J. 1993. Climatic Change: A New Security Challenge for the Atoll States of the South Pacific. *Journal of Commonwealth and Comparative Politics 31* (2): 173–192.

———. 2003. Losing Ground? Tuvalu, the Greenhouse Effect and the Garbage Can. *Asia Pacific Viewpoint 44* (2): 89–107.

———. 2010. From Blackbirds to Guestworkers in the South Pacific. *Plus ça Change…? The Economic and Labour Relations Review: ELRR* 20 (2): 111–121.

———. 2011. Elephants in the Pacific? Pacific Urbanisation and Its Discontents. *Asia Pacific Viewpoint 52* (2): 121–135.

———. 2015. Vulnerable Islands: Climate Change, Tectonic Change, and Changing Livelihoods in the Western Pacific. *The Contemporary Pacific* 27 (1): 1–36.

Connell, J., and J. Lea. 2002. *Urbanisation in the Island Pacific: Towards Sustainable Development*. London/New York: Routledge.

Corbett, J. 2015. "Everybody Knows Everybody": Practising Politics in the Pacific Islands. *Democratization* 22 (1): 51–72.

Dahl, A.L., and I.L. Baumgart. 1983. *The State of the Environment in the South Pacific*. Nairobi: UNEP.

Davis, W.J. 1996. The Alliance of Small Island States (AOSIS): The International Conscience. *Asia-Pacific Magazine* 2 (May): 17–22. Available at: http://coombs.anu.edu.au/SpecialProj/APM/TXT/davis-j-02-96.html. Accessed 22 Feb 2014.

Deitelhoff, N., and L. Wallbott. 2012. Beyond Soft Balancing: Small States and Coalition-Building in the ICC and Climate Negotiations. *Cambridge Review of International Affairs 25* (3): 345–366.

Dupont, A., and G. Pearman. 2006. *Heating Up the Planet: Climate Change and Security*. Vol. 12. Sydney: Lowy Institute for International Policy.

Dyer, M. 2017. Eating Money: Narratives of Equality on Customary Land in the Context of Natural Resource Extraction in the Solomon Islands. *The Australian Journal of Anthropology* 28 (1): 88–103.

Edwards, M.J. 1996. Climate Change, Worst-Case Analysis and Ecocolonialism in the Southwest Pacific. *Global Change, Peace and Security* 8 (1): 63–80.

———. 1999. Security Implications of a Worst-Case Scenario of Climate Change in the South-West Pacific. *Australian Geographer* 30 (3): 311–330.

Farbotko, C. 2005. Tuvalu and Climate Change: Constructions of Environmental Displacement in the Sydney Morning Herald. *Geografiska Annaler: Series B, Human Geography* 87 (4): 279–293.

———. 2010. Wishful Sinking: Disappearing Islands, Climate Refugees and Cosmopolitan Experimentation. *Asia Pacific Viewpoint* 51 (1): 47–60.

Fisher, P.B. 2011. Climate Change and Human Security in Tuvalu. *Global Change, Peace and Security 23* (3): 293–313.

FRDP. 2016. *Framework for Resilient Development in the Pacific: An Integrated Approach to Climate Change and Disaster Risk Management (FRDP) 2017–2030*. Available at: http://www.forumsec.org/resources/uploads/embeds/file/Annex%201%20-%20Framework%20for%20Resilient%20Development%20in%20the%20Pacific.pdf. Accessed 12 July 2017.

Grasso, M. 2006. An Ethics-Based Climate Agreement for the South Pacific Region. *International Environmental Agreements: Politics, Law and Economics* 6 (3): 249–270.

Grydehøj, A., and I. Kelman. 2017. The Eco-Island Trap: Climate Change Mitigation and Conspicuous Sustainability. *Area* 49 (1): 106–113.

Hayward, B. 2008. Let's Talk About the Weather: De-Centering Democratic Debate About Climate Change. *Hypatia: A Journal of Feminist Philosophy* 23 (3): 79–98.

Hviding, E. 2003. Contested Rainforests, NGOs, and Projects of Desire in Solomon Islands. *International Social Science Journal* 55 (178): 539–554.

International Institute for Sustainable Development (IISD). 1994. *Summary of the UN Global Conference on The Sustainable Development of Small Island Developing States: 25 April–6 May 1994*. Available at: http://www.iisd.ca/vol08/0828000e.html. Accessed 19 Feb 2014.

IPCC. 2013. *Working Group I Contribution to the Fifth Assessment Report of the Intergovernmental Panel on Climate Change 2013: Summary for Policymakers*. Geneva: IPCC.

Kelman, I. 2014. No Change from Climate Change: Vulnerability and Small Island Developing States. *The Geographical Journal* 180 (2): 120–129.

Kelman, I., R. Stojanov, S. Khan, O.A. Gila, B. Duží, and D. Vikhrov. 2015. Viewpoint Paper. Islander Mobilities: Any Change from Climate Change? *International Journal of Global Warming* 8 (4): 584–602.

Kempf, W. 2009. A Sea of Environmental Refugees? Oceania in an Age of Climate Change. In *Form, Macht, Differenz: Motive und Felder Ethnologischen Forschens*, ed. E. Hermann, K. Klenke, and M. Dickhardt, 191–205. Göttingen: Universitätsverlag Göttingen.

Kendall, R. 2012. Climate Change as a Security Threat to the Pacific Islands. *New Zealand Journal of Environmental Law* 16: 83–116.

Lane, R., and R. McNaught. 2009. Building Gendered Adaptation to Climate Change in the Pacific. *Gender and Development* 17 (1): 67–80.

Lazrus, H. 2012. Sea Change: Island Communities and Climate Change. *Annual Review of Anthropology* 41: 285–301.

Lomborg, Bjorn. 2016. About Those Non-Disappearing Pacific Islands. *The Wall Street Journal*, October 13.

McCubbin, S., B. Smit, and T. Pearce. 2015. Where Does Climate Fit? Vulnerability to Climate Change in the Context of Multiple Stressors in Funafuti, Tuvalu. *Global Environmental Change* 30: 43–55.

Moore, E.J., and J.W. Smith. 1995. Climatic Change and Migration from Oceania: Implications for Australia, New Zealand and the United States of America. *Population and Environment* 17 (2): 105–122.

Mortreux, C., and J. Barnett. 2009. Climate Change, Migration and Adaptation in Funafuti, Tuvalu. *Global Environmental Change* 19 (1): 105–112.

Nunn, P.D. 2013. The End of the Pacific? Effects of Sea Level Rise on Pacific Island Livelihoods. *Singapore Journal of Tropical Geography* 34 (2): 143–171.

Pernetta, J.C., and P.J. Hughes. 1990. *Implications of Expected Climate Changes in the South Pacific Region: An Overview*. Nairobi: UNEP.

Roy, P., and J. Connell. 1991. Climatic Change and the Future of Atoll States. *Journal of Coastal Research* 7 (4): 1057–1075.

Rudiak-Gould, P. 2016. Climate Change Beyond the "Environmental": The Marshallese Case. In *Anthropology and Climate Change: From Actions to Transformations*, ed. S. Crate and M. Nuttall, 2nd ed., 220–227. London: Routledge.

Shibuya, E. 1997. Roaring Mice Against the Tide: The South Pacific Islands and Agenda-Building on Global Warming. *Pacific Affairs* 69 (4): 541–555.

Shie, T.R. 2007. Rising Chinese Influence in the South Pacific: Beijing's "Island Fever". *Asian Survey* 47 (2): 307–326.

Storey, D., and S. Hunter. 2010. Kiribati: An Environmental 'Perfect Storm'. *Australian Geographer* 41 (2): 167–181.

Strokirch, K.V. 2007. The Region in Review: International Issues and Events, 2005–2006. *The Contemporary Pacific* 19 (2): 552–577.

Taplin, R. 1994. International Policy on the Greenhouse Effect and the Island South Pacific. *The Pacific Review* 7 (3): 271–281.
Thornton, A., M.T. Kerslake, and T. Binns. 2010. Alienation and Obligation: Religion and Social Change in Samoa. *Asia Pacific Viewpoint* 51 (1): 1–16.
Tisdell, C. 2008. Global Warming and the Future of Pacific Island Countries. *International Journal of Social Economics* 35 (12): 889–903.
UNFCCC. 1995. Report of The Conference of The Parties on Its First Session, Berlin, 28 March to 7 April 1995. FCCC/CP/1995/7/Add.1. Available at: http://unfccc.int/resource/docs/cop1/07a01.pdf. Accessed 24 Feb 2014.
UNGA. 1994. *Report of the Global Conference on the Sustainable Development of Small Island Developing States*, United Nations General Assembly. Document A/CONF. 167/9.
UNICEF EAPRO Media Centre. 2011. *United Nations Secretary General Calls for Unity on Climate Change in Kiribati*. Available at: https://www.unicef.org/pacificislands/media_17174.html. Accessed 2 June 2013.
Wyeth, K. 2014. Escaping a Rising Tide: Sea Level Rise and Migration in Kiribati. *Asia and The Pacific Policy Studies* 1 (1): 171–185.
Yang, J. 2009. China in the South Pacific: Hegemon on the Horizon? *The Pacific Review* 22 (2): 139–158.

CHAPTER 4

Constructing Climate Security in the Pacific

Abstract Derived from the concept of environmental security, climate security has shaped understandings of climate change in the Pacific in the last decade. Climate security has also brought the future of the Pacific into discussions about regional stability, failed states, and refugee crises. However, climate security is not a singular narrative and different discourses of climate security create differing political conditions for action and resource mobilization. Two competing constructions are the focus here: climate change and conflict, and climate change and vulnerabilities. While constructing climate security as vulnerabilities offers the most promise for addressing issues faced in the region, many actors, including governments in the Pacific, draw from both of these competing discourses where necessary.

Keywords Climate security • Environmental security • Vulnerabilities • Environmental refugees

The previous chapter explored the existence of key policy narratives on climate change in the Pacific. This chapter refines and expands that analysis through a focus on the concept of climate security. Within the Pacific climate security can be located at the nexus between discourses of environmental security and policy narratives of climate change. Given the analysis of policy narratives in the previous chapter we will begin the discussion

M. Williams, D. McDuie-Ra, *Combatting Climate Change in the Pacific*, https://doi.org/10.1007/978-3-319-69647-8_4

through engagement with competing discourses of security and environmental security.

Attention to climate security as a policy problem is relatively recent in the Pacific and emerged in the wake of already competing discourses of environmental security. The concept of climate security is located in broader discourses and perspectives on security (as a generic concept) and environmental security. If we acknowledge that security is a contested concept, it follows that the search for a precise definition or singular conception of climate security will prove fruitless. To better understand the meaning of climate security therefore a brief discussion of the meta-narratives within which climate security is located is warranted. This section therefore first explores contests over the meaning of security and second, debates over the meaning of environmental security.

Security is primarily concerned with protection from threats. The various debates over the meaning of security are essentially concerned with two dimensions of security—security from what? And security for whom? (Booth 1991). The first dimension seeks to identify the source of the threat or vulnerability warranting security. The second dimension refers to the referent object of security, that is, what or who is being threatened and requires security.

The conceptual landscape of competing approaches to security can be classified according to a three-fold typology (O'Brien and Williams 2010: 398–402). First, in the traditional state-centric approach to security the source of threat is principally focused on violence and military conflict, and the referent object of security is the nation-state. This traditional approach dominated security studies during the era of the Cold War. Second, the umbrella term 'new security studies' is used to denote various approaches to security that challenged either the nature of the source of the threat or the referent object of security. Some of these new approaches to security were developed during the Cold War but challenges to the traditional paradigm intensified after the fall of the Berlin Wall. Contributions from critical security approaches and feminist scholars were important in documenting the limitations of the traditional approach. In terms of the perception of the nature and source of threat new security studies' analysts shifted from an exclusive focus on military conflict and inter-state violence to examine challenges from economic, environmental, political, and social sectors (Buzan 1991). In a telling and important move new security studies' perspectives developed a critique of the state-centric focus of the traditional approach through recognition that securing the state need not lead to security for specific individuals within the national polity. The third perspective on security is denoted by the term human security. The referent object of security in the human security approach is the individual thus marking a significant departure from the traditional state-centric approach. Moreover, in the human security approach there is a distinctive construction of the source and nature of threat. In geographical terms the source of threat is now located internally as well as externally. And a twin focus on 'freedom from fear' and 'freedom from want' extends the source and nature of the existential threat to economic, food, health, environmental, personal, community, and political dimensions of security (UNDP 1994: 24–25).

Environmental security emerged from the new and extended agenda of security studies after the end of the Cold War. It has been argued that the agenda of environmental security encompasses seven issues: 'efforts to redefine security; theories about environmental factors in violent conflict; the environmental security of the nation; the linkages between the military and environmental issues; the ecological security agenda; the environmental security of people; and the issue of securitisation' (Barnett 2001: 8). A decade after this observation the relationship between environment and security and the meaning of environmental security remains the subject of debate. As Elliott has demonstrated:

> there are at least nine sometimes overlapping and sometimes mutually exclusive views about the ways in which environmental degradation, resource depletion and the ecological crisis are either connected to security issues or, more specifically, can be defined as threats to security. (2007: 41)

The key issues are: resource decline and environmental degradation as the source of actual conflict between states; environmental degradation as a cause of diplomatic disputes, political tension, or the deployment of military capability short of prolonged conflict; environmental degradation as a contribution to the breakdown of societal relations, especially in fragile states; resources and environmental services as a 'weapon' of war; environmental degradation as consequence of war and conflict; environmental threats to economic security; environmental degradation linked to other forms of non-traditional security; environmental security and human security; security of and for the environment (ibid: 41–46).

Despite this overabundance of linkages between security and the environment the various cleavages can usefully be conceived as giving rise to two discourses on security and the environment (Detraz 2011; Elliott 2007). Detraz (2011: 106–107) distinguishes between an environmental conflict discourse which links environmental issues with traditional security concerns and an environmental security discourse with a focus on vulnerabilities experienced by human beings as a result of environmental degradation. In a similar manner, Elliott (2007: 137–138) distinguishes between a 'modified realist position', which focuses on the role of environmental degradation in producing armed conflict and thus threatening the security of states, and a 'human security position' in which the focus is on the role of environmental degradation in jeopardizing sustainable livelihoods and thus threatening the security of individuals and communities. We agree with these authors that there are indeed two discursive constructions of environmental security. As noted above one approach has its antecedents in the traditional security paradigm and is primarily focused on inter-state conflict and the likelihood that a military response will be required in the face of security challenges raised by environmental degradation. An alternative approach owes its intellectual heritage to non-traditional perspectives on security (including human security) and frames the impact of environmental degradation in terms of vulnerabilities (Webersik 2010) and the appropriate security response in the context of resilience and cooperative behaviour. Greater attention is given to the impact of environmental degradation and resulting insecurities for individuals and communities rather than for states (McDonald 2013). Here we refer to these two approaches to environmental security simply as the 'conflict approach' and the 'vulnerabilities approach'. While not wishing to add further conceptual vocabulary to the debate merely for the sake of it, we believe that these terms capture the broad intent of both Detraz and Elliot while also marking a clearer distinction between the two different approaches.

Given the existence of competing perspectives on environmental security it is important to acknowledge that experiences and understandings of environmental security cannot be assumed, they differ within states, national groups, ethnic groups, and communities. Below we will illustrate the ways in which differing framings of environmental security lead to different constructions of climate security in the Pacific. But prior to that exercise it is necessary to discuss briefly the process through which different perspectives on security seek to become authoritative.

The concept of securitization introduced by the Copenhagen School provides a useful tool with which to understand the existence of competing security discourses. Security as has been argued above is not an objective state of affairs but arises from social processes of interaction in which neither the source of threat nor the referent object is a given. In other words there is nothing inherently natural about either threats or referent objects; rather these are produced through social processes (Williams 2003). For the Copenhagen School any issue can be securitized, that is, presented as a security threat and thus requiring responses commensurate with the level of threat perception. The focus moves from an objective definition of security to understanding security as an inter-subjective process. In short, 'the exact definition and criteria of securitization is constituted by the intersubjective establishment of an existential threat with a saliency sufficient to have substantial political effects' (Buzan et al. 1998: 25). Securitization thus requires the acceptance by the relevant target audience of the construction of threat made by authoritative actors (Balzacq 2005).

It has been argued that the framing of climate change as a security issue 'raises the issue of how security institutions deals with climate change and what kind of preparations are being discussed to deal with various scenarios' (Dalby 2009: 143). However, while largely accurate this position is also limiting since it assumes a singular approach to the securitization of climate change, that is, one in which security institutions are involved. In contradistinction to Dalby we note *competing* and *parallel* approaches to the framing of climate change as a security issue, not all of which require the engagement of security institutions. In our understanding securitization of an issue is a distinctly political and social process in which the outcome is not predetermined. Indeed, we argue that climate security in the Pacific is an unfinished security project in which competing regionalist and securitizing discourses can be discerned. In the rest of this chapter we examine two competing securitization discourses around climate change in the Pacific.

Climate Change and Conflict

In this section, we examine the securitization of climate security as threat. We do this through first establishing the core thesis of environmental security conceptualized in terms of violence and conflict, and second, through a focus on the agents and processes framing climate security in this way in the Pacific. We argue that in the Pacific climate change is securitized by external actors and by the governments of PICTs themselves, both in response to the discursive construction presented by international actors and donors, and as a way of moving climate change and the Pacific region further up the global agenda.

Climate security has been grafted on to existing discourses of environmental security. Framing environmental security (and hence climate security) in terms of conflict processes rests on four narratives (Detraz 2011: 109–112). In the first narrative, environmental degradation leads to conflict over diminishing natural resources (see Homer-Dixon 1991; Kaplan 1994). This is the central thesis at the heart of the environmental security literature. Conflicts are predicted on a number of axes: between nation states, over shared water sources, within nation states, over land, and between communities moving within and across international borders. Further concerns about negative impacts to global agricultural and fisheries production, food security, and development are seen as potential triggers to armed conflict. A second and related narrative concerns climate-induced migration, instability, and violence. The pressures on natural resources will fuel migration that has the potential to lead to further conflict, usually discussed in terms of environmental refugees (Reuveny 2007). In addition, conflict itself will lead to further refugee flows. Inherent in this story is a concern with state capacity to deal with these conditions. Different states will experience the impacts of climate change in different ways, yet different states will also have highly variable resources to contend with these impacts (Barnett and Adger 2007). A third narrative draws attention to the security in already volatile regions containing states considered fragile or failing, including sub-Saharan Africa (Brown et al. 2007; Raleigh 2010), the Middle East, parts of Asia, and the Pacific. In this scenario climate change induces state collapse with catastrophic consequences for the regions themselves and the international system (Parsons 2011). More extreme realist scholars have made links between climate change, state collapse, and terrorism (Smith 2007). The final narrative embedded in this conflict-prone approach to environmental security is a focus on national security. In other words, the referent object of study is the nation-state and threats are perceived to be important only in so far as they impact on the core objectives of states.

As the environmental security thesis framed in terms of conflict has been applied to climate change it has also come under deeper critical analysis. Most critical approaches cite a lack of empirical evidence for claims made about the inevitability of conflict arising from climate change. For instance, Salehyan (2008) refutes the 'deterministic' logic at the heart of the thesis and calls for more careful and complex analysis of the links between resources and violence. Hendrix and Glaser (2007) have analysed the likelihood of civil conflicts from climate change related impacts in sub-Saharan Africa and argue that climate change impacts are not necessarily more likely to trigger conflict than other factors. Theisen (2008) makes a similar point, arguing that empirical evidence shows that land degradation increases the likelihood of conflict but that land degradation takes place with and without considering climate change. Hsiang et al. (2011) have argued that it is extreme weather conditions, not long-term climate change, that correlates with increased conflict.

Despite these critiques the thesis remains prominent at the global level (Detraz and Betsill 2009). At the instigation of the British government, the United Nations Security Council (UNSC) held its first debate on the impact of climate change on peace and security in 2007. While some delegates disputed the competence of the Security Council to debate climate change arguing that it is an issue of sustainable development and not peace and security. Margaret Beckett, the then UK Foreign Secretary in introducing the debate relied on an expanded view of security in which climate change was 'about collective security in an increasingly fragile world for all' and not simply an issue of 'narrow national security'. But she maintained that it was a 'threat multiplier'; and asserted that 'a full account of climate risks should be undertaken when examining the root causes of conflict' (United Nations Security Council 2007). In July 2011 at the conclusion of a one-day debate on climate change and security the President of the Security Council issued a statement that noted that 'The Security Council expresses its concern that possible adverse effects of climate change may, in the long run, aggravate certain existing threats to international peace and security' (United Nations Security Council 2011). The Security Council held another meeting in 2015 on *Climate Change as a Threat Multiplier to Global Security.* Of note was the speech made by Robert Aisi, the ambassador to the UN from Papua New Guinea speaking on behalf of all PICTs with UN representation. Aisi noted: 'We have heard how climate change will have profound effects on food production, water availability, territorial integrity, human migration, and forced displacement—all of which are potentially destabilizing—yet we have only the vaguest idea of what an international response to these growing impacts would look like' (Aisi 2015: n.p). Aisi's comments are telling of the acceptance of the links between climate change and security threats on the one hand and the challenge of action to address these impacts at the international level.

The links between climate change and international peace and security have been made in policy-making circles and in academic writing (Scott 2015). There are two main ways in which the traditional security paradigm is used to construct climate change in the Pacific. The first is the creation of environmental refugees or climate refugees seen as undermining stability in the states where they are created and in the states receiving them. The second is the impacts of scarcity on state stability and the potential for fragile PICTs to fragment further into failed states. Recent cases of instability in Fiji (Bellamy 2008) and the Solomon Islands (Wainwright 2003) act as a portent for a region where instability is the norm and regional powers such as Australia will have to deal with the consequences. Edwards (1996) argues that the conditions for military conflict exist in the Pacific, but until the threat of climate change there has been no catalyst for conflict. While vague, the underlying argument assumes that armed conflict could lead desperate states to request military assistance from different sources, inviting a proxy theatre of conflict drawing in China, Taiwan, and the United States and its allies, including Australia. This converges with deeper concerns about Chinese and Taiwanese competition in the Pacific and with China's increased involvement with PICTs as a donor and investor (Atkinson 2007; Yang 2009).

An official discourse linking climate change with conflict and traditional security concerns is discernible in the statements of several regional governments. In this perspective climate change will further instability, violence, and refugee flows in the Pacific threatening already fragile states and undermining regional security in the broader Asia-Pacific. The link between climate security and conflict is grafted onto an existing construction of security challenges in the Pacific arising from state failure and fragility. While explicit attention to climate change as a source of insecurity is relatively recent a focus on governance failure and economic weakness as a security threat in specific Pacific countries has been echoed in a number of Australian government publications (see, e.g. Australian Government 2003: 20; Australian Government 2005: 916). Such views were not confined to the Department of Defence but also Australia's official aid agency (now part of the Department of Foreign Affairs and Trade) which noted in its *Pacific Regional Aid Strategy 2004–2009* that 'a porous and undeveloped region is not in the interests of the Pacific or Australia' (AusAid 2007: 3). From this perspective unless this insecurity is addressed it will generate problems that directly impact on Australian economic, foreign, and security policies. The Government of New Zealand expressed similar views at the Pacific Islands Forum in Niue in 2008 (Podesta and Ogden 2008).

Kendall (2012: 99) and Dupont and Pearman (2006) refer to IPCC reports and comments made by the United Nations to demonstrate that climate-induced migration will likely cause competition over fewer resources and deepen internal divisions and divisions between states. Based on these sources, they argue that due to sea level rises Pacific Islanders will first move inland, then eventually to other countries, mainly Australia and New Zealand, which could cause both national and regional insecurity. Locke (2009) also, through a qualitative approach and case study of Kiribati and Tuvalu, illustrates that movements from rural and outer islands to urban central islands is related to a combination of the impacts of climate change and socio-economic factors and, whilst he does not identify the national insecurity due to urban migration, he does illustrate that international migration from the PICTs to developed countries may pose a security risk. Locke (2009: 177–178) points to the Australian magazine, *Security Solutions* that suggests that climate change is a national security threat as forced migration can be linked to terrorism. Locke (2009), however, argues that this security risk may be minimized if the right policies that address climate-induced migration are applied.

Some of the literature also builds on the security concern that the effects of climate change on the PICTs will also have impacts upon other nations in various ways, especially in relation to environmental refugees and that there must be a national policy of adaptation to cope with this eventuality. This is further supported by Koser (2012) who argues that it is in Australia's national interest to have a policy on climate change refugees and to work on adaptation strategies with the PICTs, due to Australia's strategic, economic, and development relations with the region.

Perhaps the most definitive example of constructing climate change as a security threat in the Pacific is the Australian Department of Defence's *White Paper 2009* (Australian Government, Department of Defence 2009). In the *White Paper* climate change was recognized as a key issue shaping Australia's strategic outlook for the first time. Notwithstanding an existing approach to climate security issues in the Pacific that focus on vulnerability and emphasize the necessity of developing resilience and capabilities (AusAid 2007; Australian Government 2009) the *White Paper* constructs climate change simply in terms of the conflict perspective. In a discussion of 'new security concerns' identified as climate change and resource security the authors of the paper are explicit in constructing climate security through the lens of conflict. For instance, the paper states:

> The Government also considered new security risks that might arise from the potential impact of climate change and resource security issues, involving future tensions over the supply of energy, food and water. These issues are likely to exacerbate already significant population, infrastructure and governance problems in developing countries, straining their capacity to adapt. (ibid: 39)

Furthermore, although recognizing the increasing vulnerability of weak and fragile states from the effects of climate change the conflict frame is dominant. The *White Paper* further asserts:

> ...the new potential sources of conflict related to our planet's changing climate, or resource scarcity, give rise to very old forms of confrontation and war, such as clashes between states over resources. From a defence planning point of view, the key issue concerns the nature of such conflicts and the implications for defence capabilities, rather than their cause (ibid: 40).

This framing of climate security through the lens of conflict and linking it to traditional security concerns is not the preserve of the dominant regional power but has also been expressed by leaders from PICTs and representatives in major international venues. For example, during the debate in the UN Security Council on climate change and peace and security in 2007 the representative from Papua New Guinea speaking on behalf of the PIF claimed that 'the impact of climate change on small islands was no less threatening than the dangers guns and bombs posed to large nations' (United Nations Security Council 2007). Moreover, in a speech to the 63rd United Nations General Assembly (UNGA) in September 2008, Emanuel Mori, President of the Federated States of Micronesia asserted that 'climate change ... impacts international peace and our own security, territorial integrity and our very existence, as inhabitants of the very small and vulnerable island nations' (cited in Oxfam 2009: 19). And a year later also at the UN General Assembly, the representative of Nauru speaking on behalf of the Pacific Small Island Developing States, claimed that 'resettlement and migration were already occurring and dangers to international peace and security' (United Nations Department of Public Information 2009). This conflict frame was echoed by Palau's representative who noted that 'We do not carelessly call climate change a security threat. When we are told by scientists to prepare for humanitarian crisis, including exodus, in our lifetimes, how can it be different from preparing for a threat like war?' (ibid).

Other governments in other forums have echoed similar sentiments. Constructing climate change as conflict follows the logic of securitization articulated by the Copenhagen School. Climate change *becomes* a security issue through the ways social and political processes frame the issue. Once it is framed through the conflict lens alternative approaches become more difficult to articulate and when articulated are subsumed by the dominant discourse. While it is tempting to view the discourse solely as a product of global and regional actors, as noted above, actors representing PICTs also reproduce the discourse. There are several ways of interpreting this reproduction. First, reproducing climate as conflict discourse is a strategic manoeuvre by PICTs to gain the attention of donors and international actors in order to access resources to combat climate change and to gain legitimacy and bargaining power in international climate negotiations. Second, reproducing the discourse gives legitimacy to PICT governments among their own populations and affirms the state as the primary actor in securing the population against climate-induced conflict. Third, governments of PICTs are convinced by the arguments of international and regional actors that continually emphasize the links between climate change and violent conflict and thus express a sincere anxiety. Fourth, governments of PICTs may also reproduce the climate as conflict discourse because to deny it or downplay it risks jeopardizing resources, legitimacy, and domestic support. These interpretations are not mutually exclusive and it is reasonable to assume that in some PICTs they converge to some extent.

Climate Change and Vulnerabilities

An alternative perspective to a focus on conflict is the construction of climate change primarily in terms of vulnerability and through developing linkages with the non-traditional security agenda and the human security framework. A human security frame has been explicitly deployed by academics (see Elliott 2010; McDonald 2013) and civil society organizations (Oxfam 2009) but unlike the conflict perspective, the framing of vulnerabilities as a security issue is often implicit rather than explicit in official discourse. Environmental insecurity refers here to the insecurities experienced by individuals, communities, and in turn, states through a range of interlinked environmental issues including, though not restricted to, deforestation, access to natural resources, marine and coastal degradation, and climate change. From this perspective it is possible to focus on the relationships between climate change and other forms of insecurity. Literature focussing on the impacts of climate change in developing countries identifies various links between climate change and a variety of socio-economic factors. We argue that framing environmental security (and hence climate security) in terms of vulnerabilities rests on six key narratives. In addition to the four narratives (loss of livelihood, worsening poverty, food security, and human health) delineated by Detraz (2011: 112–114), we add two additional narratives, namely gender and migration.

One key narrative which focuses on vulnerabilities is that linking climate change and health. It is widely agreed that climate change constitutes a global health problem (McMichael et al. 2006) although researchers note that the linkages between health and climate change will have regional specificities (Haines et al. 2006). Researchers are not agreed on the ways in which to specify the multifaceted linkages between climate change and health. We can delineate six vectors through which the health impacts of climate change can be studied: changing patterns of disease and morbidity, water and sanitation, food, population and migration, health system infrastructure, and extreme weather events. The vulnerability of populations in the Pacific Island countries have increased and are likely to be affected as a result of the impact climate change on health. First, there is evidence that climate change is responsible for an increase in the frequency of tropical diseases, water-borne diseases, contamination, and diarrhoeal illness (Russell 2011). For example, the number of recorded cases of malaria in Papua New Guinea's Western Highlands Province increased from 638 in 2000 to 4986 in 2005 (Oxfam 2009: 18). Second, existing poor sanitation, water pollution and limited access to fresh drinking water (coupled with poor hygienic practices) is likely to be further damaged as a result of climate change, for example through saline intrusion into ground water tables with a consequent rise in communicable diseases such as cholera and gastroenteric diseases (Russell 2011: 10). Third, food security is widely recognized as a major threat to public health in many Pacific countries. One consequence of decreasing local food productivity has been the rise in imported food of dubious nutritional value. Such lifestyle changes have led to increases in 'stroke, hypertension, heart disease, and type 2 diabetes' (ibid: 11). Fourth, migration and urbanization has placed strains on existing health infrastructure, and on water, weak and sewage systems. Furthermore, urbanization concentrates vulnerable populations in small areas where diseases are more easily able to spread. Fifth, the limited health care infrastructure of Pacific countries increases their vulnerability to the impacts of climate change. It has been argued that 'the remote geography of Pacific SIDS [small island developing countries] ... creates numerous challenges for their health systems; amongst these is the retention and training of medical staff, the collection of health system data and the provision of infrastructure' (Lovell 2011: 52). Finally, the impact of extreme weather events is both direct in terms of heat waves and morbidity, for example, and indirect in relation to damage to health services. Examples of indirect effects in the Pacific include the severe impact of cyclone Heta in 2004 on the only hospital in Niue and the loss of hospital equipment in Papua New Guinea as a result of flooding in December 2008 (Oxfam 2009: 18).

A second narrative links climate change and food security. The literature suggests that climate change will compromise food security in three main ways. First, climate variation reduces crop yields and destroys entire crops (Brown and Funk 2008). This links to the health concerns outlined above, as imported food of poor nutritional value replaces local crops. New pests and diseases will harm humans, plants, and livestock thus compromising food safety and food security (FAO n.d.). Second, the impact of climate change on oceans, seas, lakes, and rivers will adversely impact those whose livelihood depends on fishing and aquaculture. It has been estimated that around 200 million people globally will be affected (ibid). This is a major source of vulnerability in the Pacific where fish is a staple of local diets. Third, climate change will increase hunger and malnutrition through a reduction in agricultural yields, loss of forest products and decreasing fisheries (ibid). Populations directly dependent on food production for food security are especially at risk and dependency on imported food will further increase the vulnerability of local populations to further climate related ruptures to food elsewhere and increases the costs of food.

In an influential report the Food and Agricultural Organization (FAO 2008) documented the links between climate change and food security in the Pacific. As a direct result of variations in the normal rainfall pattern consequent on climate change the report argues that there will be a disruption to agriculture production, which is heavily dependent on summer rainfall. Moreover, sea level rise will lead to attendant problems such as salinization, coastal erosion and inundation which may threaten food security by reducing available agricultural land. The impact on agriculture will be replicated in fisheries with sea level rise and sea surface temperature changes leading to a decline in fisheries productivity. Since the consumption of fish in Pacific Island countries is high (at around 70 kilograms per person per year) the adverse impact of climate change on fisheries will negatively impact food security. For PICTs these developments compromises local food production and increases dependency on foreign imports (Barnett 2010); increases the marginalization of people in outlying and remote parts of the region who will be even more vulnerable to food availability and price rises/variations. As Barnett and Adger (2003: 322–3) demonstrate, atoll countries are even more vulnerable to food insecurity than other parts of the Pacific.

A third narrative links climate change with poverty. The foreword to a major Organization of Economic Cooperation and Development (OECD) report claims that 'Climate change is a serious risk to poverty reduction and threatens to undo decades of development' (OECD n.d.: V). Although the OECD report is not framed in the language of human security it provides compelling evidence of the negative impact of worsening poverty on human security. The authors detail the ways in which worsening poverty undermines health, food security, ecosystems, population movements, and livelihoods. Running throughout the report is an emphasis on vulnerabilities. The authors acknowledge that because climate change is 'superimposed on existing vulnerabilities '(ibid: IX) and therefore 'in general, the vulnerability is highest for least developed countries in the tropical and subtropical areas. Hence, the countries with the fewest resources are likely to bear the greatest burden of climate change in terms of loss of life and relative effect on investment and the economy' (ibid: X)

A fourth narrative links climate change to loss of livelihood. Worsening levels of poverty in PICTs is a result of adverse changes to livelihoods. Declining agricultural production, threats to the fishing industry, and ruptures to the tourism industry, all threaten livelihoods. It is important to note that supplementary livelihoods that provide crucial cash income for poor households, female-headed households, and populations in remote areas will also be threatened along with more conventional livelihoods measurable through unemployment and underemployment figures. This will place increased pressure on remittances for livelihoods. The notion of declining industries may also shape perceptions of risk in new ways. For instance, the idea that livelihoods in a particular area may not have long-term viability may accelerate environmental degradation of say, timber resources, in order to make the most revenue in a short period (Barnett and Adger 2003). Furthermore, the notion of impending catastrophe in the Pacific can affect investment in PICTs and prevent a long-term outlook on development planning by government and donors (Mortreux and Barnett 2009). Degradation to ecosystems from climate change threatens livelihoods through reducing the natural resource base of many local economies and also placing increasing pressure on less vulnerable land and resources. Thus as coastal land is under threat in PICTs, there is an increase in pressure on inland areas, forests, and minerals accelerating degradation in these areas. Degradation adds further push factors to migration, including to urban areas, less affected islands, other parts of the region, and beyond. As Barnett has argued, 'this makes migration an attractive if not the only option to preserve livelihoods and quality of life' (2003: 12). Migration options are dependent on neighbouring countries. Australia and New Zealand have been discussed as possible destinations for a larger group of migrants from the Pacific for some time but migration within PICTs and between them is also placing increased pressure on livelihoods, land, and housing (Locke 2009).

In addition to the four narratives noted above we also document two further narratives linking climate security with vulnerabilities. First, while migration is dominant in the framing of climate security as conflict viewing migration through the lens of vulnerability steers the issue away from the potential for conflict in host communities and the environmental refugees debate towards current migration dynamics and the varied spaces within which migration takes place. Throughout PICTs migration is becoming an increasingly important part of livelihoods, whether it is migration within PICTs, including from outer islands to larger island and from rural to urban areas, migration between PICTs, and migration outside PICTs as guest workers and skilled migrants. Migration in all these spaces engenders insecurity for those migrating, those remaining behind, and those competing in the same labour markets. In some locations internal migration exacerbates existing tensions between ethnic groups, as in the Solomon Islands (Dinnen 2002), while in other locations like Tuvalu, decisions to stay rather than migrate demonstrate the complexities of local notions of insecurity (Mortreux and Barnett 2009). Crucially a focus on vulnerability allows for a more differentiated view of migration dynamics taking place now (rather than potentially) and their relationship to insecurity. Localized community and individual perceptions of risk, the role of government and civil society in constructing insecurity at the local level, and the experience of migration in community narratives all become more visible through this approach. In other words, migration out of PICTs is no longer an inevitable outcome of climate change, but is based on individual and community decisions in the context of these changes.

Second, gender relations are potentially reconfigured under the impacts of climate change (Denton 2002). The relationship between gender and climate change is co-constitutive. A focus on vulnerability brings gender-related security issues to the fore. And a focus on gender brings vulnerabilities to the fore (Detraz 2009: 354). Changes in food production and access, livelihoods, migration, and household structure all affect gender relations. In PICTs there are further issues to consider. Women are poorly represented in political decision-making, though in some locations they have a strong voice in civil society. Levels of gender-based violence are high and may be exacerbated by ruptures to livelihoods, households, and increased migration. Women and men are forced to adapt to new circumstances, yet these changes take place in social contexts that are not necessarily changing as rapidly and continue to have highly gendered expectations around labour, familial responsibilities, and voice. In addition, climate aid and adaptation measures have gendered impacts in the same ways as other development aid.

Implications of the Competing Discourses

As the discussion of climate security framed in terms of vulnerabilities demonstrates, constructing climate change as non-traditional security allows for climate change to be analysed in four distinct ways to the dominant modes of climate as conflict. First, it enables a reconsideration of space. More specifically it invites analysis of the sub-national dimensions of climate change impacts. Going beyond a national level analysis renders visible the differentiated impacts of climate change within different nation states. The small land area and population size of many PICTs makes sub-national analysis uncommon. Furthermore there is a tendency to homogenize PICTs into a shared category awaiting a shared fate. However, the immense geographic, ethnic, and economic diversity of the Pacific region and particular PICTs means that the insecurities emanating from climate change need to be considered on a different scale to most orthodox analyses.

Second, a vulnerabilities lens enables the impacts of climate change adaptation measures to be considered more thoroughly. This shifts the policy response from the amount of funding to analysing the impacts climate funding is having and the ways it may be better directed in the future. In other contexts development aid, policies, and agencies are analysed in immense critical depth. Yet climate adaptation measures require similar analysis at the local level. Certain actors stand to benefit from increased climate financing such as certain government ministries, contractors, or civil society actors. Their interests need to be considered in the ways climate change issues are framed and politicized through bilateral and international funding mechanisms. At present analysis of adaptation appears preoccupied with the amounts of aid flows and the capacity and uptake by national governments. Yet the impacts of adaptation measures on different communities and stakeholders and the ways in which adaptation measures are negotiated and contested at the sub-national levels needs far greater attention (Adger et al. 2003: 192). Who has a voice at the sub-national level? Who is marginalized? At present there is an implicit sentiment that any funding for climate change going towards the most vulnerable parts of the world, especially PICTs, is good funding. Yet this overlooks the opportunities for corruption and patronage within PICTs and the ways these are challenged and/or accepted domestically.

Third, a focus on vulnerabilities counter-intuitively complicates the notion of the Pacific as a vulnerable region and PICTs as vulnerable states containing vulnerable people. As discussed in the previous chapter, for Farbotko (2010) and others the Pacific has become an 'island laboratory' for testing the truth and urgency of climate change. To be sure, the Pacific is experiencing extreme insecurity from climate change. However, a non-traditional approach allows for consideration of the ways different communities and individuals adapt to the impacts of climate change through resilience. As sub-national studies of climate change in the developing world have shown, for many communities at the sub-national level, climate change is part of a more immediate set of ruptures to everyday existence. In other words, climate change is not necessarily 'special' in the way that global and national actors view it, but it is another source of insecurity that threatens livelihoods and produces hardships (Thomas and Twyman 2005: 121). To be clear, climate change is a major cause of insecurity for populations in PICTs. However, for these populations climate change-induced insecurity is not necessarily separable from a broader sense of insecurity derived from poverty, environmental degradation, and low levels of human development. Viewing climate change as part of a constitutive insecurity experienced by vulnerable populations within PICTs exposes a disjuncture between the experience of insecurity at the sub-national level and the ways climate change is constructed nationally and globally. Many communities have adapted to catastrophic ruptures before and have developed resilience. While national action plans attempt to prepare communities in PICTs to cope with change, existing coping strategies are being overlooked. However, we are conscious that romanticizing community resilience has its own problems including bypassing local needs on the basis that the community will simply cope with change.

Fourth, constructing climate change as non-traditional security may enable a radical rethinking of the development agenda in PICTs and other developing countries. Climate change has become increasingly mainstreamed in development thinking and practice, though as Grist (2008) argues, this has stopped far short of suggesting any transformative change. She reiterates that adaptation has become a by-product of doing development well (2008: 793). Boyd et al. (2009) argue that there is potential for climate change to alter the ways development is practised and that development methods may enable a clearer idea of resilience and adaptation in different developing contexts. Cannon and Müller-Mahn (2010) argue that considerations of climate change are having an impact on development discourses, but that the focus on adaptation poses a threat to the poor in developing countries as it favours resilience over vulnerability. Others argue that attention to adaptation could help reduce chronic poverty. Tanner and Mitchell (2008) argue that a poverty-centred adaptation agenda is possible and with high-level commitments to developing countries to deal with climate change, it may be possible to more systematically address chronic poverty.

In PICTs the situation is highly uneven and further research needs to be undertaken into the ways in which climate change is shaping and being shaped by existing development practices. Initial impressions suggest that climate change has overshadowed development funding and projects in much of the region as noted in Chap. 3. This has two potential impacts. First, insecurities that cannot be easily related to climate change may remain on the periphery of local development agendas. In other words, as funding for climate change related vulnerabilities increases, the ability of local actors, whether state or non-state, to draw attention to other issues will decline. Second, state and non-state actors have responded by framing vulnerabilities and insecurities as climate change related in order to fund projects. This may have the effect of drawing more resources and attention to insecurities previously neglected by governments and donors, but it also means that grassroots actors have to navigate new levels of bureaucracy to fund development.

Conclusion

Competing discourses of climate security configure and reconfigure the PICTs in various future scenarios from violent conflict to state collapse. The dominant discourse grafts climate change onto orthodox approaches to environmental security. This discourse has been used to construct climate change in the Pacific by international actors, particularly UN agencies, and by regional powers, especially Australia. And also, at times, by governments in PICTs themselves to pursue resources, legitimacy, state primacy, and perhaps reflect some belief in the security scenario at the heart of the discourse. In contrast, constructing climate change through a focus on vulnerabilities links climate change to the production and reproduction of chronic insecurity for populations within PICTs. We have outlined six narratives linking climate change to vulnerabilities. Using this vulnerabilities approach enables crucial departures from the conflict approach. Specifically, it enables a sub-national focus, draws attention to the impacts of existing measures, recognizes the mutually constitutive impacts of climate change vulnerabilities and other vulnerabilities experienced by individuals and communities in PICTs, and offers new possibilities for rethinking development practice and priorities in the region. By comparing these approaches it is clear that a focus on vulnerabilities challenges the emerging orthodoxy on climate change and security and environmental security more broadly. However, given the saliency of the emerging orthodoxy within and outside the region, constructing climate change through vulnerabilities faces significant obstacles. Yet increasingly the two approaches coexist and at times compete with one another. Many actors, including PICT governments, draw from both when necessary. Further research needs to consider the conditions under which the different discourses are politically possible and the pathways for action emanating from each.

References

Adger, W.N., S. Huq, K. Brown, D. Conway, and M. Hulme. 2003. Adaptation to Climate Change in the Developing World. *Progress in Development Studies* 3 (3): 179–195.

Aisi, R. 2015. *Statement for the UN Security Council, Open Arria-Formula Meeting on the Role of Climate Change as a Threat Multiplier for Global Security.* Available at: http://www.spainun.org/wp-content/uploads/2015/07/Papua-New-Guinea_CC_201506.pdf. Accessed 20 June 2017.

Atkinson, J. 2007. Vanuatu in Australia-China-Taiwan Relations. *Australian Journal of International Affairs* 61 (3): 351–366.

AusAid. 2007. *Aid and the Environment—Building Resilience, Sustaining Growth: An Environment Strategy for Australian Aid.* Canberra: Commonwealth of Australia.

Australian Government. 2009. *Engaging Our Pacific Neighbours on Climate Change: Australia's Approach.* Canberra: Commonwealth of Australia.

Australian Government, Department of Defence. 2003. *Australia's National Security: A Defence Update 2003.* Canberra: Commonwealth of Australia.

———. 2005. *Australia's National Security: A Defence Update 2005.* Canberra: Commonwealth of Australia.

———. 2009. *Defending Australia in the Asia Pacific Century: Force 30. Defence White Paper 2009.* Canberra: Commonwealth of Australia.

Balzacq, T. 2005. The Three Faces of Securitization: Political Agency, Audience and Context. *European Journal of International Relations* 11 (2): 171–201.

Barnett, J. 2001. Adapting to Climate Change in Pacific Island Countries: The Problem of Uncertainty. *World Development 29* (6): 977–993.

———. 2003. Security and Climate Change. *Global Environmental Change 13* (1): 7–17.

———. 2010. Dangerous Climate Change in the Pacific Islands: Food Production and Food Security. *Regional Environmental Change* 11 (1): 229–237.

Barnett, J., and W.N. Adger. 2003. Climate Dangers and Atoll Countries. *Climatic Change* 61 (3): 321–337.

———. 2007. Climate Change, Human Security and Violent Conflict. *Political Geography* 26 (6): 639–655.

Bellamy, P. 2008. The 2006 Fiji Coup and Impact on Human Security. *Journal of Human Security* 4 (2): 4–18.

Booth, K. 1991. Security in Anarchy: Utopian Realism in Theory and Practice. *International Affairs* 67 (3): 527–545.

Boyd, E., N. Grist, S. Juhola, and V. Nelson. 2009. Exploring Development Futures in a Changing Climate: Frontiers for Development Policy and Practice. *Development Policy Review* 47 (6): 659–674.

Brown, M.E., and C. Funk. 2008. *Food Security Under Climate Change*. National Aeronautics and Space Administration. University of Nebraska–Lincoln Paper 131. Available at: http://digitalcommons.unl.edu. Accessed 12 July 2015.

Brown, O., A. Hammill, and R. McLeman. 2007. Climate Change as the 'New' Security Threat: Implications for Africa. *International Affairs* 83 (6): 1141–1154.

Buzan, B. 1991. New Patterns of Global Security in the Twenty-First Century. *International Affairs 67* (3): 431–451. Waever.

Buzan, B., O. Weaver, and J. de Wilde. 1998. *Security: A New Framework for Analysis*. Boulder/London: Lynne Rienner.

Cannon, T., and D. Müller-Mahn. 2010. Vulnerability, Resilience and Development Discourses in Context of Climate Change. *Natural Hazards* 55 (3): 621–635.

Dalby, S. 2009. *Security and Environmental Change*. Cambridge: Polity.

Denton, F. 2002. Climate Change Vulnerability, Impacts, and Adaptation: Why Does Gender Matter? *Gender and Development* 10 (2): 10–20.

Detraz, N. 2009. Environmental Security and Gender: Necessary Shifts in an Evolving Debate. *Security Studies* 18 (2): 345–369.

———. 2011. Threats or Vulnerabilities? Assessing the Link Between Climate Change and Security. *Global Environmental Politics* 11 (3): 104–120.

Detraz, N., and M.M. Betsill. 2009. Climate Change and Environmental Security: For Whom the Discourse Shifts. *International Studies Perspectives* 10 (3): 303–320.

Dinnen, S. 2002. Winners and Losers: Politics and Disorder in the Solomon Islands 2000–2002. *The Journal of Pacific History* 37 (2): 285–298.

Dupont, A., and G. Pearman. 2006. *Heating Up the Planet: Climate Change and Security*. Vol. 12. Sydney: Lowy Institute for International Policy.

Edwards, M.J. 1996. Climate Change, Worst-Case Analysis and Ecocolonialism in the Southwest Pacific. *Global Change, Peace and Security* 8 (1): 63–80.

Elliott, L. 2007. Environment and Security: What's the Connection? *Australian Defence Force Journal* 174: 39–52.

———. 2010. Climate Migration and Climate Migrants: What Threat, Whose Security? In *Climate Change and Displacement: Multidisciplinary Perspectives*, ed. Jane McAdam, 175–190. Oxford/Portland: Hart Publishing.

FAO. n.d. Climate Change and Food Security. Available at: http://www.fao.org/climatechange. Accessed 23 Sep 2016.

Farbotko, C. 2010. Wishful Sinking: Disappearing Islands, Climate Refugees and Cosmopolitan Experimentation. *Asia Pacific Viewpoint* 51 (1): 47–60.

FAO. 2008. *Climate Change and Food Security in Pacific Island Countries*. Rome: Food and Agricultural Organization.

Grist, N. 2008. Positioning Climate Change in Sustainable Development Discourse. *Journal of International Development* 20 (6): 783–803.

Haines, A., R.S. Kovats, D. Campbell-Lendrum, and C. Corvalan. 2006. Climate Change and Human Health: Impacts, Vulnerability, and Mitigation. *Lancet* 367 (9528): 2101–2109.

Hendrix, C.S., and S.M. Glaser. 2007. Trends and Triggers: Climate, Climate Change and Civil Conflict in Sub-Saharan Africa. *Political Geography 26* (6): 695–715.

Homer-Dixon, T.F. 1991. On the Threshold: Environmental Changes as Causes of Acute Conflict. *International Security* 16 (2): 76–116.

Hsiang, S.M., K.C. Meng, and M.A. Cane. 2011. Civil Conflicts Are Associated with the Global Climate. *Nature* 476 (7361): 438–441.

Kaplan, R.D. 1994. The Coming Anarchy. *The Atlantic Monthly* 273 (February): 44–76.

Kendall, R. 2012. Climate Change as a Security Threat to the Pacific Islands. *New Zealand Journal of Environmental Law* 16: 83–116.

Koser, K. 2012. *Environmental Change and Migration: Implications for Australia.* Sydney: Lowy Institute for International Policy. http://www.lowyinstitute.org/publications/environmental-change-and-migration-implications-australia. Accessed 12 July 2017.

Locke, J.T. 2009. Climate Change-Induced Migration in the Pacific Region: Sudden Crisis and Long-Term Developments. *The Geographical Journal* 175 (3): 171–180.

Lovell, S.A. 2011. Health Governance and the Impact of Climate Change on Pacific Small-Island Developing States. *IHDP Update* 1: 50–55.

McDonald, M. 2013. Discourses of Climate Security. *Political Geography* 33: 42–51.

McMichael, A., R. Woodruff, and S. Hales. 2006. Climate Change and Human Health: Present and Future Risks. *The Lancet* 367 (9513): 859–869.

Mortreux, C., and J. Barnett. 2009. Climate Change, Migration and Adaptation in Funafuti, Tuvalu. *Global Environmental Change* 19 (1): 105–112.

O'Brien, R., and M. Williams. 2010. *Global Political Economy: Evolution and Dynamics.* 3rd ed. Houndmills: Palgrave Macmillan.

OECD. n.d. *Poverty and Climate Change: Reducing the Vulnerability of the Poor Through Adaptation.* Available at: http://www.oecd.org/dataoecd/60/27/2502872.pdf. Accessed 12 Mar 2014.

Oxfam. 2009. *The Future Is Here: Climate Change in the Pacific.* Carlton/Newton: Oxfam Australia/Oxfam New Zealand.

Parsons, R.J. 2011. Strengthening Sovereignty: Security and Sustainability in an Era of Climate Change. *Sustainability* 3 (9): 1416–1451.

Podesta, J., and P. Ogden. 2008. The Security Implications of Climate Change. *Washington Quarterly* 31 (1): 115–138.

Raleigh, C. 2010. Political Marginalization, Climate Change, and Conflict in African Sahel States. *International Studies Review* 12 (1): 69–86.

Reuveny, R. 2007. Climate Change-Induced Migration and Violent Conflict. *Political Geography* 26 (6): 656–673.

Russell, L. 2011. *Poverty, Climate Change and Health in Pacific Island Countries: Issues to Consider in Discussion, Debate and Policy Development.* Sydney: University of Sydney and Australian National University, Menzies Centre for Health and Public Policy.

Salehyan, I. 2008. From Climate Change to Conflict? No Consensus Yet. *Journal of Peace Research* 45 (3): 315–326.

Scott, S.V. 2015. Implications of Climate Change for the UN Security Council: Mapping the Range of Potential Policy Responses. *International Affairs* 91 (6): 1317–1333.

Smith, P.J. 2007. Climate Change, Weak States and the 'War on Terrorism' in South and Southeast Asia. *Contemporary Southeast Asia* 29 (2): 264–285.

Tanner, T., and T. Mitchell. 2008. Entrenchment or Enhancement: Could Climate Change Adaptation Help to Reduce Chronic Poverty? *IDS Bulletin* 39 (4): 6–15.

Theisen, O.M. 2008. Blood and Soil? Resource Scarcity and Internal Armed Conflict Revisited. *Journal of Peace Research* 45 (6): 801–818.

Thomas, D.S., and C. Twyman. 2005. Equity and Justice in Climate Change Adaptation Amongst Natural-Resource-Dependent Societies. *Global Environmental Change* 15 (2): 115–124.

UNDP. 1994. *Human Development Report.* Geneva: UNDP.

United Nations Department of Public Information. 2009. General Assembly, Expressing Deep Concern, Invites Major United Nations Organs to Intensify Efforts in Addressing Security implications of Climate Change. News Release GA/10830 3 June.

United Nations Security Council. 2007. *Security Council Holds First Ever Debate on Impact of Climate Change on Peace, Security Hearing Over 50 Speakers.* UN Security Council SC/9000.

———. 2011. *Statement by the President of the Council.* S/PRST/2011/15.

Wainwright, E. 2003. Responding to State Failure: The Case of Australia and the Solomon Islands. *Australian Journal of International Affairs* 57 (3): 485–498.

Webersik, C. 2010. *Climate Change and Security: A Gathering Storm of Global Challenges.* Santa Barbara: Praeger.

Williams, M.C. 2003. Words, Images, Enemies: Securitization and International Politics. *International Studies Quarterly* 47 (4): 511–531.

Yang, J. 2009. China in the South Pacific: Hegemon on the Horizon? *The Pacific Review* 22 (2): 139–158.

CHAPTER 5

Organizing Climate Finance in the Pacific

Abstract The demand for climate finance by Pacific states to combat climate change is derived from their inability to finance adaptation and mitigation projects from their own resources. However, climate finance is donor-driven and is entwined with development assistance, often using the same language, frameworks, and norms. The chapter begins by examining the global architecture of climate finance. The Pacific is a subset of this global architecture and operates under many of the same mechanisms and norms. The second part engages with the institutional framework for climate finance in the Pacific. The third part explores the normative implications of an emerging climate finance regime in the Pacific and the impact of a fragmented, diverse, and complicated regime in the region where it matters most of all.

Keywords Climate finance • Aid dependency • Regional climate finance • Climate finance in the Pacific

Financial flows to the Pacific targeted at mitigation and adaptation efforts are governed by norms, practices, and decisions primarily developed by donor nations. The first part of the chapter therefore examines the global architecture of climate finance. It documents the sources, delivery mechanisms, and institutional framework for climate finance. In the second part of the chapter attention is focused on the institutional framework for

M. Williams, D. McDuie-Ra, *Combatting Climate Change in the Pacific*, https://doi.org/10.1007/978-3-319-69647-8_5

climate finance in the Pacific. It examines bilateral and multilateral funding initiatives, the role of regional organizations in coordinating the distribution of funds, and the sectoral allocation of funding. The third part of the chapter explores the implications of an emerging climate finance regime in the Pacific. A combination of complex funding structures, inadequate sources of finance, and the absence of donor coordination reduces the effectiveness of climate finance in the Pacific. We assess the role of regional organizations in supporting the efforts of individual states to maximize access to financial resources. In the absence of an agreed definition of climate finance, the term will be understood in this chapter to refer to financial flows mobilized by developed countries to support climate change mitigation and adaptation measures in developing countries.

The Global Architecture of Climate Finance

In the search for financial assistance to combat climate change Pacific Island nations face an evolving and complex landscape of climate finance. The global architecture of climate finance is fragmented, diverse, and complicated (Amerasinghe et al. 2017; Nakhooda et al. 2016). If the existence of a regime produces stability of expectations and therefore contributes to enhanced effective governance through the creation of rules, norms, and agreed decision-making procedures, climate finance can be construed as existing in a pre-regime phase. Of course, regimes are dynamic and can be the site of conflict and contention. However, while the existence of a regime is not synonymous with stability its absence tends to indicate an absence of stable expectations.

As Fig. 5.1 shows financial flows for climate change adaptation and mitigation is a complex process involving several actors. The focus of this chapter will be on finance provided through the UNFCCC process, bilateral finance institutions, and multilateral finance institutions.

While the subject of climate finance is an outcome of the UNFCCC process it is not confined to the UNFCCC with negotiations on climate finance taking place in a variety of institutions. The High-Level Advisory Group on Climate Change Financing established by the then UN Secretary General, Ban Ki-Moon in 2010 highlighted four sources of climate finance (2010): public sources, development bank instruments, market-based finance, and private capital. As is evident from Fig. 5.2 the approach to climate finance has followed that of development assistance with the

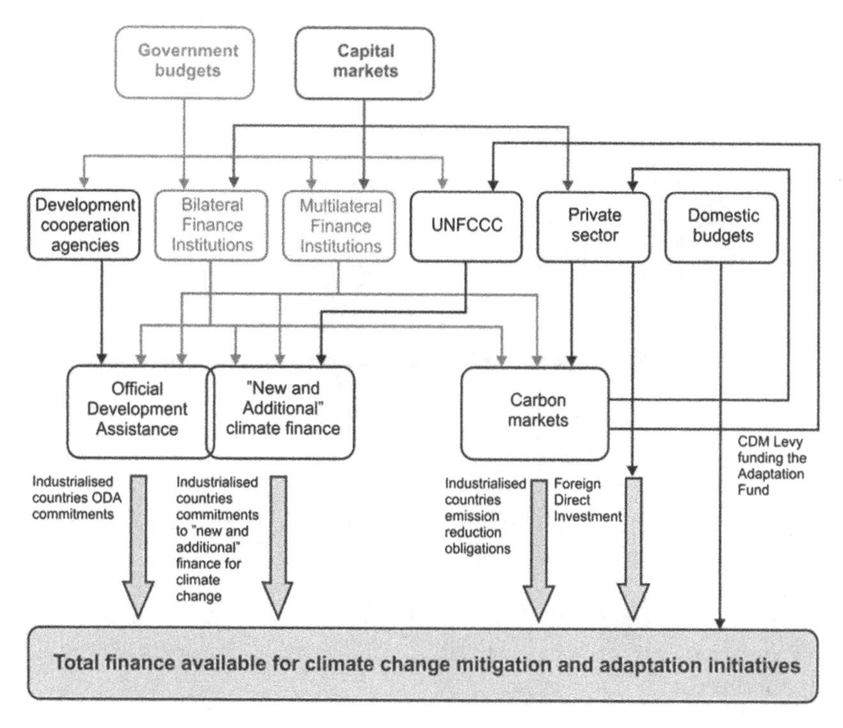

Fig. 5.1 Financial flows for climate change mitigation and adaptation in developing countries (Source: Atterridge et al. 2009)

greatest concentration coming from bilateral sources. Governments tend to prefer bilateral allocation since this gives them greater control over the distribution of funds.

It is well established that the provision of funding for public programmes can be sourced from bilateral, multilateral, or private agents (Schalatek and Bird 2016). In the absence of an overarching global framework the institutions of climate finance have created a patchwork of agencies giving rise to a fragmented, diverse, and complicated landscape. In itself the existence of multiple sources of finance need not create a governance problem. However, the absence of clear mandates, defined jurisdictions, and transparent procedures exacerbate the difficulties of coordinating a system in which multiple donors fund adaptation and mitigation projects. This has resulted in a lack of coordination, duplication of effort, and the entrenchment of inefficiency. Moreover, beginning with

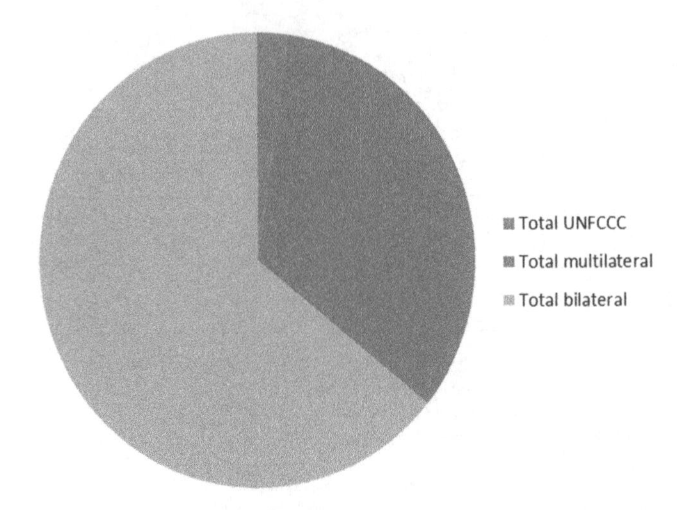

Fig. 5.2 Global climate finance pledges according to the type of source (Source: Extracted from Climate Funds Update Data)

the Kyoto Protocol that foregrounded a market-based approach the focus has been on market-based mechanisms and private sector finance is an important part of the regime.

Multilateral Sources of Climate Finance

There are currently 22 climate funds in operation with multilateral funding the dominant source (see Table 5.1).

The two most significant funding mechanisms are the Global Environmental Facility (GEF) and the Green Climate Fund (GCF). The GEF was established in 1991 as a three-way partnership involving the UNEP, UNDP, and the World Bank for administering UNFCCC funds as well as key multilateral funds that have not been created directly through decisions of the COP. The stated mission of the GEF is to serve as 'a mechanism for international cooperation for the purpose of providing new and additional, grant and concessional funding to meet the agreed incremental costs of measures to achieve agreed global environmental benefits' (Ehlers 2011). The GEF provides financing to country-driven climate change mitigation and climate change adaptation projects. Funding for GEF projects have risen slowly from $2 billion allocated to GEF-1 in 1994 to $4.43 billion dedicated to funding under GEF-6 (2014–2018). GEF funding is

Table 5.1 Sources of climate finance (funds, administrating bodies, organizational type, and stated purpose)

Name	*Administrative body*	*Organizational type*	*Stated purpose*
Adaptation Fund	Adaptation Fund Board	Multilateral	Adaptation
Adaptation for Smallholder Agriculture	The International Fund for Agricultural Development (IFAD)		Adaptation
Amazon Fund	Brazilian Development Bank (BNDES)	Multi-donor National	Mitigation—REDD
Bicarbon Fund	The World Bank	Multilateral	Adaptation, Mitigation—general, Mitigation—REDD
Clean Technology Fund	The World Bank	Multilateral	Mitigation—general
Congo Basin Forest Fund	African Development Bank	Multi-donor Regional	Mitigation—REDD
Forest Carbon Partnership Facility	The World Bank	Multilateral	Mitigation—REDD
Forest Investment Program	The World Bank	Multilateral	Mitigation—REDD
GEF Trust Fund—Climate Change focal area	The Global Environment Facility (GEF)	Multilateral	Adaptation, Mitigation—genera
Global Climate Change Alliance	The European Commission	Multilateral	Adaptation, Mitigation—general, Mitigation—REDD
Global Energy Efficiency and Renewable Energy Fund	The European Commission	Multilateral	Mitigation—general
Green Climate Fund	GCF	Multilateral	Adaptation, Mitigation—general, Mitigation—REDD
Indonesia Climate Change Trust Fund	Indonesia's National Development Planning Agency	Multi-donor National	Adaptation, Mitigation—general, Mitigation—REDD
Least Developed Countries Fund	The Global Environment Facility (GEF)	Multilateral	Adaptation
MDG Achievement Fund—Environment and Climate Change thematic window	UNDP	Multilateral	Adaptation, Mitigation—general

(*continued*)

Table 5.1 (continued)

Name	*Administrative body*	*Organizational type*	*Stated purpose*
Partnership for Market Readiness	The World Bank	Multilateral	Mitigation—general
Pilot Program for Climate Resilience	The World Bank	Multilateral	Adaptation
Scaling-Up Renewable Energy Program for Low Income Countries	The World Bank	Multilateral	Mitigation—general
Special Climate Change Fund	The Global Environment Facility (GEF)	Multilateral	Adaptation
Strategic Climate Fund	The World Bank	Multilateral	Adaptation, Mitigation—general, Mitigation—REDD
Strategic Priority on Adaptation	The Global Environment Facility (GEF)	Multilateral	Adaptation
UN-REDD Programme	UNDP	Multilateral	Mitigation—REDD

Source: Climate Funds Update.

heavily weighted towards mitigation projects. As at 30 June 2017, the GEF has provided $5.3 billion in funding for 867 mitigation projects on climate change mechanism (CCM), and $1.17 billion in grant funding for adaptation projects (GEF 2017: 1–4). Adaptation funding is administered through the Least Developed Countries Fund (LDCF), the Special Climate Fund (SCCF), and the Strategic Priority on Adaptation (SPA).

The Green Climate Fund (GCF) was established at COP 16 in Cancun as an operating entity of the financial mechanism of the Convention under Article 11 (Decision 1/COP16), and launched at COP 17 in 2011 in Durban (Decision3/COP 17). The headquarters of the Fund is in Songdo, Incheon, Korea. The GCF is the largest multilateral climate fund and is expected to become the primary channel through which international public climate finance will flow over time. It aims to adopt a country-driven approach, to balance adaptation and mitigation finance, allow direct access, and have a private sector facility (Schalatek et al. 2016).

Table 5.2 Multilateral Funds Supporting Adaptation, 2003–2013 (US$ millions)

Fund	*Pledged*	*Deposited*	*Approved*	*Projects approved*
Adaptation for Smallholder Agriculture Program	336.45	336.25	345.00	48
Adaptation Fund	569.15	546.91	348.91	54
Green Climate Fund	10,255.39	9,896.38	324.19	14
Least Developed Countries Fund (LDCF)	1,250.16	1,077.01	973.24	237
Pilot Program for Climate Resilience (PPCR)	1,117.00	1,117.00	972.50	65
Special Climate Change Fund (SCCF)	367.31	362.31	298.50	71

Source: Caravani et al. (2016): 1

Initial funding for the GCF was provided through pledges by contributing countries. By mid-2016, 43 countries had pledged $10.3 billion. The GCF became operational in 2015 and has since disbursed $2.2 billion (for 43 projects). An evolving financial institution the GCF has 58 projects and programmes in its pipeline, seeking $3.4 billion in funding (Green Climate Fund 2017).

As is evident from Table 5.1 climate change funding flows to both adaptation (building resilience of communities to impacts) and mitigation (reducing carbon pollution). From the perspective of Pacific nations most attention is focused on adaptation funding (Caravani et al. 2016). The main funds contributing to supporting adaptation efforts are detailed in Table 5.2.

Bilateral Sources of Climate Finance

Bilateral channels are an important source of climate finance. At the 15th COP in Copenhagen in 2009 developed country signatories of the UNFCCC committed to achieving a goal of mobilizing $100 billion a year by 2020 to support mitigation and adaptation activities in developing countries. Twenty-two countries and the European Commission give bilateral funding for climate change. Among the major bilateral contributors to climate finance are Germany's International Climate Initiative, the UK's International Climate Fund, Australia's International Forest Carbon Initiative, and Norway's International Forest Climate Initiative.

Concomitant on the emergence of various sources of funds has been political contestation over the objectives of climate finance and the required volume of finance. These debates are compounded by the absence of an agreed definition of climate finance. For example, the UNFCCC's Standing Committee on Finance (2014: 5) defines climate finance in terms of its objectives.

> Climate finance aims at reducing emissions, and enhancing sinks of greenhouse gases and aims at reducing vulnerability of, and maintaining and increasing the resilience of, human and ecological systems to negative climate change impacts.

In contradistinction a UNEP report (2011: 5) defined climate finance in terms of a donor-recipient model linked to specific goals:

> Climate finance is finance flowing from developed to developing countries, including support for mitigation, adaptation, and related policy and capacity-building.

The consequence of a failure to agree on a consensual definition is a situation in which the objectives of climate finance are not self-evident and the interests of 'donors' and 'recipients' have not always been aligned with different actors pursuing competing normative agendas. The multilateral banks and bilateral agencies instrumental in providing climate funding have attempted to establish norms governing climate finance that reflect their interests as major funders. On the other hand, national and regional organizations seeking finance have tended to emphasize issues related to accessibility, adequacy, and coordination. In other words, recipient need is defined in terms of greater accessibility of funds, adequate amounts, and coordination of climate funds. Civil society organizations have also intervened in climate finance debates urging greater volumes of public finance, increased equity, and enhanced transparency.

One key contention has concerned the relative weight given to finance for mitigation (see Patel et al. 2016) as against finance for adaptation (see Caravani et al. 2016). The needs of small island developing countries centre around finance for adaptation, although they do receive finance for mitigation efforts (for further discussion see Durand et al. 2015; Watson et al. 2016b). The discussion above is not meant to indicate a set of arrangements with multilateral funding taking place separately from bilateral efforts. Schemes such as REDD+ (see Watson et al. 2016a) frequently involve bilateral and multilateral funding; public and private sources of climate finance.

Moreover, negotiations on climate finance emerged in the context of sustainable development, and donor agencies have grafted climate finance on to the pre-existing regime of official development assistance. Arguably, the inherent link between poverty alleviation, sustainable development, and climate change has changed the concept of official development assistance, expanding its traditional focus from economic development and welfare to include environmental sustainability and protection from catastrophic climate change threats. The linkages between climate finance and official development assistance have been made at the conceptual and operational levels. At the conceptual level it has been recognized that the effects of climate change will have a severe impact on the world's poorest populations. Around the world, millions of poor people are already at risk of tragic crop failures, reduced agricultural productivity, increased malnutrition and hunger, water scarcity, and the spread of infectious diseases. For example, the *World Development Report 2010* estimates that the developing countries will bear between 75% and 80% of the costs of damages associated with climate change (World Bank 2009). From the perspective of governments and multilateral agencies attempts to address the negative consequences of climate change can be perceived as a global public good with two principal aspects, mitigation and adaptation. First, efforts to mitigate climate change help to ensure long-term sustainable development for the entire global community, in both developed and developing countries. Second, adaptation assistance is critical in protecting the world's poorest people from potentially devastating climate change effects. Thus, the liberal rationale for foreign aid as both ensuring greater security for the international community and at the same time alleviating poverty in the developing world secures a normative basis for climate finance.

Another key issue in the debate over climate finance centres on the volume of finance required to address climate change adaptation and mitigation. Unsurprisingly, given the complex nature of the issue no consensus has been reached on the adequate amount of climate finance. This failure stymies efforts to find an adequate solution to the pressing problems faced by developing countries, since an effective framework for climate finance governance requires some agreement on the volume of finance required to address climate change adaptation and mitigation. While governments and international organizations agree on the necessity of climate finance there is no consensus on the size of the resource gaps that such finance is targeted to fill. Estimates of the total volume of climate change finance required have varied widely. For example, a World Bank study estimated that adaptation would cost between $9 billion and $41 billion from 2007 to 2030 (World Bank 2006). In contrast, the Stern Review estimated adaptation will cost $4–$37 billion per annum (Stern 2007). The UNDP's *Human Development Report 2007* raised the stakes much higher with its prediction that adaptation funds needed by 2015 would be US$86–$109 billion per year (UNDP 2007). UNEP's Adaptation Gap Report (2016) has suggested an estimate of $280–$500 billion per year by 2050. The difficulty of arriving at reliable estimates and therefore consensual knowledge was summed up in the UNFCCC's report on the National Economic, Environment and Development Study for Climate Change Project. The report concluded that:

> For most of the countries engaged in the NEEDS (National Economic, Environment and Development Study for Climate Change Project) project, assessing the costs of mitigation and adaptation measures was a challenge, owing to institutional and methodological constraints. While tools and methodologies are widely available for estimating the costs of mitigation measures, they are rather scarce in relation to adaptation. One of the specific challenges is that most of the measures identified are not only for the purposes of adapting to climate change but also have other development-related benefits. (UNFCCC 2010: 6–9)

Central to current debates on the adequacy of funding is the so-called Copenhagen Accord. At the conclusion of COP 15 in December 2009 it was agreed that new and additional funding was required. The Copenhagen Accord is a non-binding document but the commitment to achieve a target of $100 billion in climate finance by 2020 has been widely accepted. This target became the key focus of discussions concerning the adequacy of climate finance in the period leading to the Paris Conference in December 2015. Prior to the Paris Conference the OECD published a report that claimed that countries were making progress in meeting the Copenhagen Accord target. According to the OECD the total climate finance flows rose to $52 billion in 2013 and further to $62 billion in 2014 (OECD 2015). The significance of the $100 billion target was further reinforced in the Paris Agreement that urges the developed countries to increase their level of support to ensure that they meet the target of providing $100 billion annually by 2020 for mitigation and adaptation.

The Institutional Framework of Climate Finance in the Pacific

It is within this complex and fragmented system that Pacific nations have to seek climate finance essential for combatting climate change as they lack the requisite financial resources to devise and implement adaptation and mitigation strategies. While partly arising from the size and scale of the challenges posed by climate change, the dependence of Pacific nations on climate finance is a result of their general economic weakness and vulnerability, and dependence on foreign aid for standard development purposes. Irrespective of the success of their current development model it is apparent that in the absence of the requisite financial capacity to invest in adaptation and mitigation policies, the nations of the Pacific are dependent on climate finance.

As argued above the context of climate finance in the Pacific has been established at the global level. Thus, the donors of climate finance to the Pacific reflect this complex and fragmented landscape. Unlike the global trend in which bilateral funding is the major source of climate finance in the Pacific, multilateral sources are dominant (see Fig. 5.3).

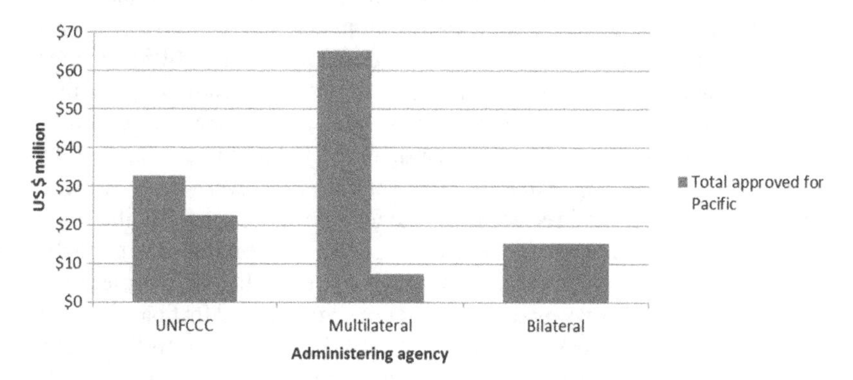

Fig. 5.3 Approved and disbursed climate finance in Pacific by type of administering agency (Source: Based on data from Climate Funds Update)

The major bilateral donors are Australia, the European Union, France, Germany, Japan, New Zealand, Norway, the United Arab Emirates, the United Kingdom, and the United States. Of the 22 climate funds in operation (see Table 5.1), 19 are of relevance to the Pacific. Three funds—the Indonesia Climate Change Trust Fund, the Amazon Fund, and the Congo Basin Forest Fund—are regionally specific in their operation. To date PICTs have received pledges from all 19 funds. Major funding has been provided via the GCF, the Adaptation Fund, the LDCF, the SCCF, the GEF Trust Fund, the Climate Investment Fund, the Pilot Program for Climate Resilience and the Forest Carbon Partnership Facility, and the UN-REDD programme.

The contested nature of the definition of climate finance and the absence of transparency makes it difficult to provide a comprehensive listing of climate-financed projects in the region. The *Samoa Power Sector Expansion Project* is illustrative of the difficulties involved. This example illustrates the caution with which statistics on climate funding has to be approached. The Japanese government reported to the OECD that its funding of the *Samoa Power Sector Expansion Project* is climate change focused. This project was reported by the Japan International Cooperation Agency (JICA) as constituting $39.4 million spent on climate change in 2007. However, the status of this project as a climate change project is unclear and raises the further issue of additionality in climate finance. On one hand, an Asian Development Bank (ADB) report outlines climate adaptation (rather than mitigation) as a special feature of the project. It states: 'The Project supports the objectives of the Government's National Adaptation Program of Action (2005) through the underground transmission network cabling programme. The programme will help to reduce exposure of transmission assets to cyclones' (ADB 2007b: 9). On the other hand, project documents dated 2007 state the purpose is to expand and diversify the power sector to meet future load growth and reduce dependence on diesel imports. There was no mention of climate change in project administration documents at this time. In the absence of hard evidence it appears that this project was framed in different ways by the Samoan government and its financial partners. For example, AusAid reported to the OECD (also in 2007) that its $11.2 million pledged to the project fulfilled combined 'climate change and desertification' Rio objectives. A further $1.6 million was committed by AusAid to the project in 2009. The majority of these JICA and

AusAid funds were delivered through the ADB as loans (ADB n.d.). The ADB is providing a $26.61 million loan and a $15.39 million grant for the project, and the government-owned Electric Power Corporation will cover the balance of $12 million (ADB 2007a). The *Samoa Power Sector Expansion Project* does not seem to have begun its life as a climate change-focused project but it is framed as a climate change project by major funders. Given the sums involved the inclusion or exclusion of this single project has a major impact on the ways in which the balance between bilateral and multilateral sources of finance flowing into the Pacific can be viewed.

Priorities of Climate Change Finance

No single measure exists by which the priority of funders can be determined. For instance, if one looks at the stated purpose of climate funds pledged, there does not seem to be a distinct division between mitigation and adaptation activities. However, the result is different on a project-level basis. Figures 5.4 and 5.5 are based on data covering the period 2012–2014 (Climate Change Funds Update) on the stated purpose of each fund in

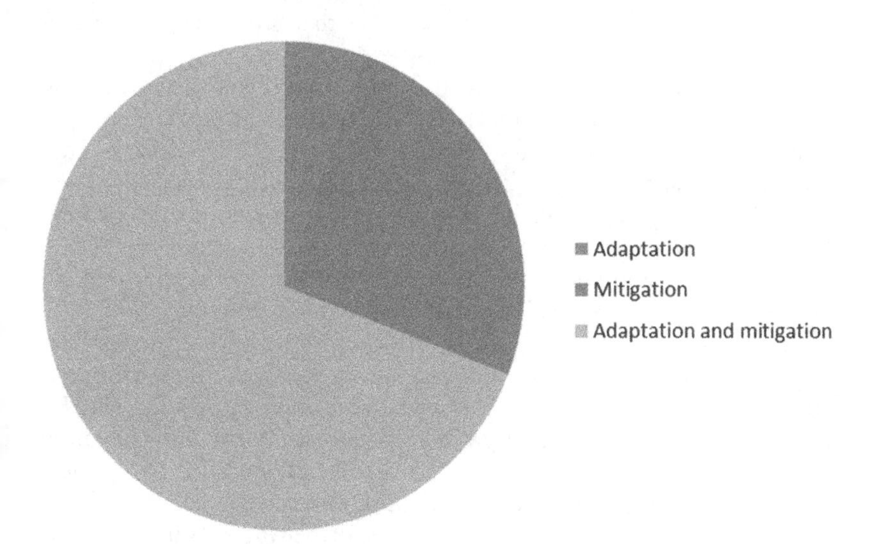

Fig. 5.4 Climate finance by stated purpose of fund

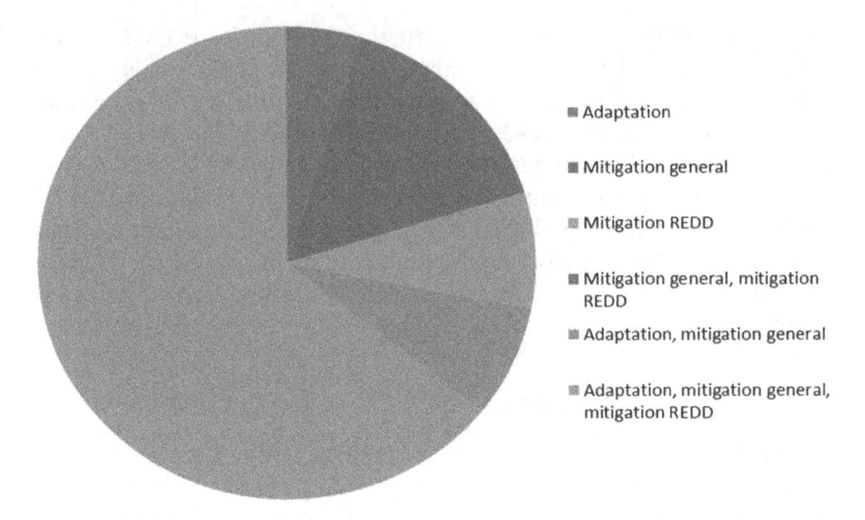

Fig. 5.5 Climate finance by focus of national pledges (adaptation, mitigation general, mitigation REDD)

aggregate terms. It does not indicate the proportion of mitigation versus adaptation-focused funds on a project level that are undertaken across categories.

An illustration of complications that arise from this way of measuring priorities is the GEF Trust Fund. The GEF Trust Fund has both mitigation and adaptation as a stated purpose. However, 98.5% of funds from the climate change focal area are spent on mitigation activities.

Looking at pledges on a project level paints a different picture. Figure 5.6 indicates a definite focus on mitigation over adaptation, REDD, and projects with multiple foci. Of the US$8.09 billion of approved climate projects, US$6.27 billion is dedicated to mitigation only.

From the perspective of PICTs adaptation measures are of greater importance than those concerned with mitigation since these countries are minor contributors to greenhouse gas emissions. However, only 15% of finance approved since 2003 have been earmarked for adaptation projects. In the Pacific the Adaptation Fund has approved one project for the Solomon Islands, and endorsed project concepts for Papua New Guinea, Fiji, and the Cook Islands. Three further projects in the Pacific have been proposed by UNDP in the Cook Islands, Samoa, and Papua New Guinea.

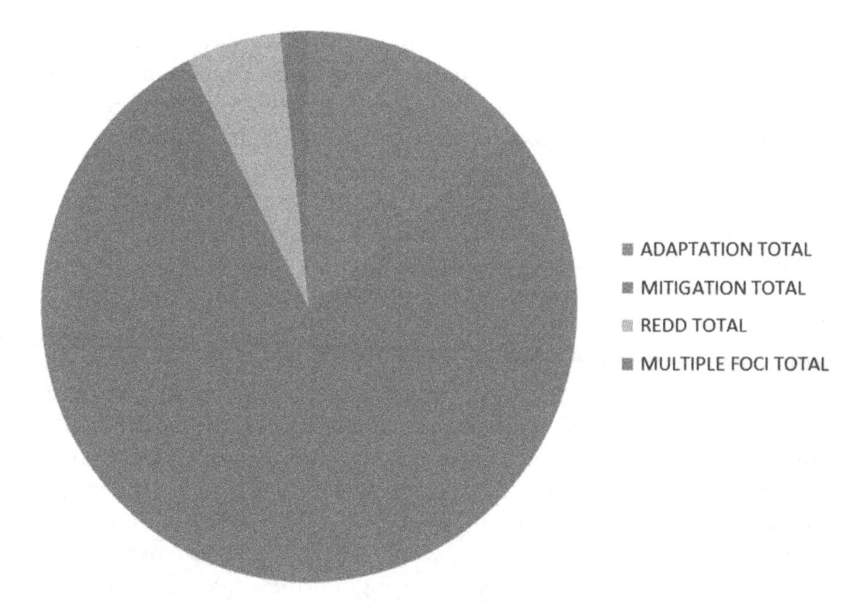

Fig. 5.6 Climate finance by approved individual projects

The Political Implications of an Emerging Climate Finance Regime in the Pacific

The third part of this chapter explores the normative implications of an emerging climate finance regime in the Pacific. It examines both the challenges identified and the political response of the PICTs.

Challenges

The climate finance regime has created a number of challenges for PICTs. A recent study (Maclellan and Meads 2016) identified three key challenges facing Pacific nations: the adequacy of climate finance given the costs facing the islands, access to climate finance, and the quality of climate finance. In this section we focus on three challenges that intersect with and mirror those discussed by Maclellan and Meads. These are a complex funding structure with multiple channels of funds, a donor-driven approach lacking in effective coordination, and the absence of country ownership. In the absence of a unified system climate change financing is channelled through a variety of different mechanisms, including direct budget support, multilateral banks, global funds, and regional organizations. This results in a funding structure that is complex, difficult to understand, and at times impossible to negotiate. As, Kosi Latu, the Director General of SPREP noted, 'Access to climate finance needs to be simplified' (Maclellan and Meads 2016: 12). These problems are not new. In an address to the United Nations General Assembly in 2010, Anote Tong from Kiribati pointedly noted that 'We regret to say that up until now we have not been able to access any of the fast-start funds pledged' (Maclellan 2011: 7). Second, uncoordinated donor action results in a complex set of donor requirements with a plethora of funding mechanisms each with their own accounting obligations, varying timelines, and complex reporting requirements. Toke Talagi from Niue expressed the frustration felt by Pacific governments when he stated, 'Each and every donor and acronym has its own governance and accountability requirements, which is challenging and frustrating for us' (Maclellan 2011: 12). This absence of donor coordination not only limits the (already weak) capacity of Pacific governments to effectively manage new flows of financing. Furthermore, the absence of country ownership inherent in these practices exacerbates the difficulty encountered in integrating climate change into national planning. This has stimulated criticism of the Adaptation Fund from a Pacific perspective. Critics have focused on the identification of vulnerabilities and have insisted that more credence should be given to the importance of local knowledge and 'country-driven' identification processes.

Responding to the Challenges of Climate Finance

Pacific nations have not simply been passive recipients of climate finance but have exercised their agency to demand changes to the existing architecture. They have voiced their criticism of the global architecture and articulated for change in the UNFCCC through utilizing their memberships of Pacific regional organizations and key developing country coalitions. PICT nations are active in a number of different fora in the UNFCCC process. They have expressed their opinions as individual nations in COP meetings, and on various climate finance boards as well as collectively through membership of state-based coalitions such as the AOSIS, the G77 and China, and the Coalition for Rainforest Nations. From the outset of the UNFCCC process, that is, at the first meeting of the Intergovernmental Negotiating Committee for a Framework

Convention on Climate Change, AOSIS, the G77, and PICT nations have been pursuing the topic of climate finance in their submissions. Central to the concerns of these groups has been access to, and transparency of, funds.

As members of the G77 the PICTs have supported the negotiating position adopted by this umbrella coalition. The G77 has have campaigned for an adequate definition of additionality in climate finance, and for new additional climate finance that does not impinge on existing multilateral and bilateral official development assistance (ODA) sources (INCFCCC 1991). In their first submission, the G77 proposed the creation of adequate, new and additional funding that does not involve reallocation of ongoing or required developmental programmes from developing countries' own resources (INCFCCC 1991: 4). They proposed a separate fund for each convention, and a separate general fund for activities not included in convention(s), that is, a financing mechanism separate from the GEF. The G77 has been critical of the GEF and from the beginning of negotiations have sought alternate arrangements to those proposed. They argued that 'the governance of the funding mechanisms should be transparent, democratic in nature with an equal voice for all parties; with access to all developing countries without any additionality; and provide for funding of activities according to the priority of the developing countries, taking into account the priorities identified in Agenda 21' (INCFCCC 1991: 5). They also criticized the accessibility of GEF-governed funds (see G77 and China submissions, INCFCCC 1994: 4; UNFCCC 1996a, b, 1997, 1998a, b, 2005). The G77 has also requested that the GEF prioritize funding for LDCs and SIDs (UNFCCC 2000).

As discussed in Chap. 3, membership of AOSIS has provided an effective platform for Pacific nations to campaign for changes to climate finance. Among the various issues identified as important by AOSIS three are particularly relevant in the context of debates on climate finance. First, the issue of additionality was raised by AOSIS during the preparatory talks. Since these early meetings AOSIS has tabled a number of proposals designed to ensure that new sources of finance are made available. For example, AOSIS proposed an 'International Insurance Pool' to provide financial insurance against the consequences of sea level rise. It was proposed that funds would be drawn from mandatory sources, and distributed to 'most vulnerable and low lying developing nations for loss and damage as a result of sea level rise' (INCFCCC 1991). The idea of an insurance facility for SIDs to meet the reparation costs of climate change-related severe weather events, sea level rise, and other climate change-related effects such as water shortages and coral bleaching resurfaced later in PICT submissions (e.g. UNFCCC 2004: 3–5).

A second issue of particular importance to Pacific nations articulated through the AOSIS framework is that of the importance of country ownership and the salience of local knowledge in defining vulnerabilities. AOSIS has championed the process of establishing vulnerabilities and determining the finance needed to support mitigation and adaptation on the basis of the reports of developing nations submitted to the UNFCCC. The 1992 UNFCCC stated that within three years of entry into force of the Convention for each individual Party, developing nations must report on (a) a national inventory of anthropogenic emissions, (b) a general description of steps taken or planned to implement the Convention, and (c) any other information relevant to achieving the aims for the Convention (Article 12, UNFCCC 1992). In the period since, of the 153 non-Annex I Parties, 137 have submitted their initial national communications, 24 their second national communications. All PICTs have submitted one national report (Samoa has submitted a second titled 'Greenhouse Gas Inventory'). In 1999, the COP5 established the Consultative Group of Experts on National Communications from Parties not included in Annex I to the Convention with the objective of improving the process of preparation of national communications from non-Annex I Parties (Decision 8/CP.5). PICT nations have supported additional requirements for developing nations to report to the COP on their vulnerabilities to climate change and their particular needs, linking this process to the need for funds to support knowledge production and project implementation. From around 1999 AOSIS begins making the case for national communications by non-Annex I nations a means to identify the vulnerabilities of Least Developed Countries and low-lying nations and establish the case for priority funding (e.g. UNFCCC 2002, 2003).

Third, Pacific states have been instrumental in shaping the AOSIS position on the necessity for reform of the institutional framework of climate finance. For example, AOSIS has been critical of the GEF Resource Allocation Framework, which has been criticized for limiting the flow of funds to SIDs for adaptation activities. An AOSIS submission to the UNFCCC in 2006 noted that only 4% of GEF single country project funding went to SIDs. On one hand, whilst enabling studies had been supported, these activities have not translated to adaptation projects. The AOSIS submission argued that 'Procedures for the Adaptation Fund should be simple, clear, avoid co-financing requirements, conditionalities and lengthy project preparation procedures. In particular, the concept of incremental costs should not be considered in the context of adaptation projects. Regional organizations should also play a critical role with regard to projects supported by the Adaptation Fund so that countries can access funding through regional organizations that have a better understanding of local conditions rather than large multilateral organizations. It is also important that lessons learned from adaptation projects be adequate documented and disseminated, particularly to other SIDS (UNFCCC 2006: 21–22). AOSIS has also argued that intergovernmental bodies should be given priority in coordinating technological innovation and dissemination. It has been argued that organizations such as the UNDP, the Small Island Developing States Network (SIDS/NET), and the Small Island Developing States Technical Assistance Program (SIDSTAP), which were started as a result of the Barbados UN Global Conference on the Sustainable Development of Small Island Developing States can fulfil these roles. AOSIS requested assistance to these bodies as a matter of priority, noting that in some cases it may be appropriate to establish new regional organizations for coordination purposes (UNFCCC 1998a: 14–15).

A focused regional approach has been undertaken through the creation of the Pacific Climate Change Finance Assessment Framework (PCCFAF) developed by the PIF Secretariat, the SPC, SPREP, and the CROP. This scheme has six dimensions, namely, funding sources, policies and plans, institutions, public financial management and expenditure, human capacity, and development effectiveness.

Conclusion

Despite playing a role in reshaping the global financial architecture addressing climate change, current climate finance models do not respond adequately to the needs of the Pacific Island states. The global architecture of climate finance is contested, and the failure to reach agreement over the objectives of climate finance (competing perspectives of donors and recipients), the required volume of finance (absence of consensus on the adequate amount of climate finance), and the governance of climate finance (an excess of competing agencies) negatively impacts the ability of small island states in the Pacific to secure adequate forms of finance in order to devise risk reduction strategies, boost their resilience, and promote sustainable development.

References

Amerasinghe, N., J. Thwaites, G. Larsen, and A. Ballesteros. 2017. *Future of the Funds: Exploring the Architecture of Multilateral Climate Finance*. Washington, DC: WRI.

ADB. 2007a. *Japan and Australia Extend $88M in Financial Assistance to Samoa Power Sector*. Asia Development Bank. http://www.adb.org/media/printer.asp?articleID=12273. Accessed 15 Aug 2016.

———. 2007b. *ADB, Japan and Australia Extend $88M in Financial Assistance to Samoa Power Sector*. Available at: http://www.adb.org/media/printer.asp?articleID=12273. Accessed 30 Aug 2013.

———. n.d. *Project Information Document 38183: Power Sector Project* (Formerly Supporting Energy Sector Reforms). Asia Development Bank. http://www.adb.org/projects/project.asp?id=38183. Accessed 15 Aug 2016.

Atterridge, A., C.K. Siebert, R.J.T. Klein, C. Butler, and P. Tellar. 2009. *Bilateral Finance Institutions and Climate Change: A Mapping of Climate Portfolios*. Stockholm: Stockholm Environment Institute.

Caravani, A., C. Watson, and L. Schalatek. 2016. *Climate Finance Thematic Briefing: Adaptation Finance*. London/Washington, DC: ODI and Heinrich Böll Stiftung.

Durand, A., L. Schalatek, and C. Watson. 2015. *Climate Finance Briefing: Small Island Developing States*. London/Washington, DC: ODI and Heinrich Böll Stiftung.

Ehlers, W. 2011. *Institutional Structure of the GEF*. GEF Expanded Constituency Workshop, Global Environment Facility, March 1–3, 2011, Belize City, Belize. http://www.thegef.org/gef/node/4233. Accessed 15 Aug 2016.

Green Climate Fund. 2017. *GCF in Numbers*. http://www.greenclimate.fund/documents/20182/24871/GCF_in_Numbers.pdf/226fc825-3c56-4d71-9a4c-60fd83e5fb03. Accessed 12 July 2017.

GEF. 2017. Report of the Global Environment Facility to the Twenty-third Session of the Conference of the Parties to the United Nations Framework Convention on Climate Change. Available at: https://www.thegef.org/sites/default/files/documents/Final%20COP%2023%20Report%20August%203.pdf. Accessed 10 Aug 2017.

INCFCCC. 1991. *Statement on Commitments Submitted by the Delegation of Vanuatu on Behalf of the States Members of the AOSIS, in Intergovernmental Negotiating Committee for a Framework Convention on Climate Change*. A/AC.237/WG.1/L.9.

———. 1994. *Matters Relating to Arrangements for the Financial Mechanism Implementation of Article 11, Paras. 1–4, in Intergovernmental Negotiating Committee for a Framework Convention on Climate Change*. A/AC.237/MISC.40.

Maclellan, N. 2011. *Improving Access to Climate Financing for the Pacific Islands.* Sydney: Lowy Institute.

Maclellan, N., and S. Meads. 2016. *After Paris: Climate Finance in the Pacific Islands: Strengthening Collaboration, Accelerating Access and Prioritising Adaptation for Vulnerable Communities.* Auckland: Oxfam New Zealand/ Oxfam Australia.

Nakhooda, S., C. Watson, and L. Schalatek. 2016. *The Global Climate Finance Architecture.* London/Washington, DC: ODI and Heinrich Böll Stiftung.

OECD. 2015. *Climate Finance in 2013–14 and the USD 100 Billion Goal: A Report by the OECD in Collaboration with Climate Policy Initiative.* Paris: OECD Publishing. https://doi.org/10.1787/9789264249424-en.

Patel, S., C. Watson, and L. Schalatek. 2016. *Climate Finance Thematic Briefing: Mitigation Finance.* London/Washington, DC: ODI and Heinrich Böll Stiftung.

Schalatek, L., and N. Bird. 2016. *The Principles and Criteria of Public Climate Finance—A Normative Framework.* London/Washington, DC: ODI and Heinrich Böll Stiftung.

Schalatek, L., S. Nakhooda, and C. Watson. 2016. *The Green Climate Fund.* London/Washington, DC: ODI and Heinrich Böll Stiftung.

Stern, N.H. 2007. *The Economics of Climate Change: The Stern Review.* Cambridge: Cambridge University Press.

UNEP. 2011. *Bilateral/Finance Institutions and Climate Change.*http://wedocs.unep.org/bitstream/handle/20.500.11822/7976/Mapping_report_final.pdf?sequence=3&isAllowed=y. Accessed 10 Mar 2016.

———. 2016. *The Adaptation Finance Gap Report* 2016. http://www.unep.org/adaptationgapreport/sites/unep.org.adaptationgapreport/files/documents/agr2016.pdf. Accessed 20 Jan 2017.

UNDP. 2007. *Human Development Report. 2007/8: Fighting Climate Change: Human Solidarity in a Divided World.* New York: United Nations Human Development Program.

UNFCCC. 1992. *The United Nations Framework Convention on Climate Change*, in The United Nations Framework Convention on Climate Change *FCCC/ INFORMAL/84.*

UNFCCC. 2006. *Views on Specific Policies, Programme Priorities and Eligibility Criteria and Possible Arrangements for the Management of the Adaptation Fund. In Subsidiary Body for Implementation.* FCCC/SBI/2006/MISC.7. Bonn: UNFCCC.

———. 1996a. *Financial Mechanism, Memorandum of Understanding Between the Conference of the Parties and the Global Environment Facility*, in Subsidiary Body for Implementation (ed.), *FCCC/SBI/1996/L.4/Rev.1*, Geneva.

———. 1996b. *National Communications*, in Subsidiary Body for Scientific and Technological Advice and Subsidiary Body for Implementation and Ad Hoc Group on the Berlin Mandate *FCCC/AGBM/1996/MISC.1/ADD.1*, Geneva.

———. 1997. *Activities Implemented Jointly Under the Pilot Phase*, in Subsidiary Body for Scientific and Technological Advice *FCCC/SBSTA/1997/MISC.5*, Bonn.
———. 1998a. *Position Papers on Matters Before the Eighth Sessions of the Subsidiary Bodies*, in Subsidiary Body for Scientific and Technological Advice and Subsidiary Body for Implementation *FCCC/SB/1998/MISC.4*, Bonn.
———. 1998b. *Proposals on Financial Mechanism*, in Subsidiary Body for Implementation *FCCC/SBI/1998/MISC.4*, Bonn.
———. 2000. *Financial Mechanisms, Other Matters*, in Subsidiary Body for Implementation (ed.), *FCCC/SBI/2000/MISC.1*, Lyon.
———. 2002. *National Communications from Parties Not Included in Annex I to the Convention, Work of the Consultative Group of Experts*, in Subsidiary Body for Implementation *FCCC/SBI/2002/MISC.6*, New Delhi.
———. 2003. *Financial Matters Relating to Parties Not Included in Annex I to the Convention, Financial Mechanism: The Special Climate Change Fund*, in Subsidiary Body for Implementation *FCCC/SBI/2003/MISC.1*, Bonn.
———. 2004. *Exchange of Views on UNFCCC Activities Relevant to Other Intergovernmental Meetings, Submissions from Parties*, in Conference of the Parties to the United Nations Framework Convention on Climate Change (ed.), *FCCC/CP/2004/MISC.2*, Buenos Aires.
———. 2005. *Draft Decision Proposed by Jamaica on Behalf of the Group of 77 and China*, in Subsidiary Body for Implementation (ed.), *FCCC/SBI/2005/CRP.1*, Montreal.
———. 2010. *Decision 1/CP.16*. http://unfccc.int/resource/docs/2010/cop16/eng/07a01.pdf#page=2. Accessed 15 Mar 2017.
———. 2014. *2014 Biennial Assessment and Overview of Climate Finance Flows Report*. UNFCCC Standing Committee on Finance. https://unfccc.int/files/cooperation_and_support/financial_mechanism/standing_committee/application/pdf/2014_biennial_assessment_and_overview_of_climate_finance_flows_report_web.pdf. Accessed 20 Aug 2016.
World Bank. 2006. *An Investment Framework for Clean Energy and Development*. Washington, DC: World Bank.
———. 2009. *World Development Report* 2010: *Development and Climate Change*. Washington, DC: World Bank.
Watson, C., S. Patel, and L. Schalatek. 2016a. *Climate Finance Thematic Briefing: REDD+ Finance*. London/Washington, DC: ODI and Heinrich Böll Stiftung.
Watson, C., S. Patel, A. Durand, and L. Schalatek. 2016b. *Climate Finance Briefing: Small Island Developing States*. London/Washington, DC: ODI and Heinrich Böll Stiftung.

CHAPTER 6

Conclusion: The Future of Climate Politics in the Pacific

Abstract The COP 21 meeting in Paris held high hopes for global climate governance. Pacific states—among others—pushed very hard for a global commitment to 1.5 °C temperature rise. The resulting Paris Agreement fell short of expectations for Pacific states. The concluding chapter begins by analysing the implications for the Pacific following the landmark Paris Agreement and what the future of regional climate politics may look like. This is followed by a brief summary of the content and main arguments presented in the book and a consideration of future research possibilities.

Keywords COP 21 Paris • Paris Agreement • Global environmental politics • Climate insecurity

The Future of Climate Politics in the Pacific

The previous chapters in this book have demonstrated that climate security for Pacific Island nations are profoundly shaped by the global climate regime. The leaders of the PICTs repeatedly emphasized the limitations of the Kyoto regime and were prominent in the discussions on the successor global climate framework. The long and tortuous negotiations to conclude a successor to the Kyoto Protocol was finally achieved at the COP

M. Williams, D. McDuie-Ra, *Combatting Climate Change in the Pacific*, https://doi.org/10.1007/978-3-319-69647-8_6

21 in Paris in December 2015. The central aim of the Paris Agreement is to strengthen the global response to the threat of climate change. Governments agreed:

> to a long-term goal of keeping the increase in global average temperature to well below 2 °C above pre-industrial levels;
> to aim to limit the increase to 1.5 °C, since this would significantly reduce risks and the impacts of climate change;
> on the need for global emissions to peak as soon as possible, recognizing that this will take longer for developing countries;
> to undertake rapid reductions thereafter in accordance with the best available science.

The UNFCCC noted that the Paris Agreement 'builds upon the Convention' and this is obviously the case. However, there are several key differences between the Paris Agreement and the Kyoto climate regime. First, the Paris Agreement is the first-ever universal, legally binding global climate deal. With the Paris Agreement, the distinction between Annex 1 and other countries has been abolished. The Paris Agreement brings all nations together in a collective effort to combat climate change. Second, and relatedly, the rigid distinction between developed and developing countries has been superseded with a more nuanced approach. The Paris Agreement recognizes differentiation among countries within a single framework. That is, states will meet their commitments in line with their capabilities. Third, the Paris Agreement has replaced the traditional top-down approach with a bottom-up framework. Previous international agreements (e.g. like the Kyoto Protocol) established targets determined at the global level and required states to meet these targets. Under the Paris Agreement states will establish their own emissions reductions targets. These 'nationally determined contributions' are to be revised and strengthened over time. All Parties are required to report regularly on their emissions and on their implementation efforts. Fourth, the agreement to hold a process of global stocktaking every five years 'in the light of equity and best available evidence' (UNFCCC 2015) will push countries towards more ambitious targets and prevent backsliding. Fifth, the primary focus of the Kyoto Protocol was on mitigation. With the Paris Agreement there is an increased focus on adaptation. Sixth, the agreement recognizes the role of non-Party stakeholders in addressing climate change, including cities, other subnational authorities, civil society, the private sector, and others. They are invited to scale up their efforts and support actions to reduce emissions, build resilience and decrease vulnerability to the adverse effects of climate change, uphold and promote regional and international cooperation.

The Paris Agreement is unequivocally a diplomatic success. It ended almost a decade of stalemate in international climate change negotiations. The UN process is no longer stagnant and there is renewed hope that the agreement can deliver meaningful outcomes for rich states, emerging economies, and poor, island states. As Dimitrov (2016: 2) noted, 'All major protagonists endorsed the deal, and countries with diametrically opposed interests supported it.' Nevertheless, while the Paris Agreement represents a definite departure from the Kyoto regime it is unclear whether the results will be positive for the Pacific Islands. In other words, there is no certainty that the Paris Agreement will enable states and other actors to achieve the ambitious goals established. Critical analysts have argued that the Paris Agreement is not fundamentally different from the Kyoto framework since they both rely on the economic processes and systems that produced the problem in the first place. From this perspective, the environmental managerialist approach embedded in both the Kyoto and Paris documents can never lead to an effective solution to the problems caused by climate change. As Spash (2016) argues, instead of leading to cuts in greenhouse gas emissions as soon as possible, the Parties promise escalation of damages and treat worst-case scenarios as a 50:50 chance. He contends that the Paris Agreement signifies commitment to sustained industrial growth and risk management over disaster prevention, and is based on future inventions and technology as the solution. From this perspective, the Paris Agreement represents a business-as-usual scenario and this approach is incompatible with tackling greenhouse gas emission.

In September 2015 the Leaders of the Pacific Islands Development Forum (PIDF) met in Suva, Fiji, and issued the Suva Climate Declaration. The Suva Declaration lists 14 'demands' the leaders hoped would be met at the Paris summit. We will briefly review three of these 'demands'. The first was the call for the '2015 Paris Climate Change Agreement to limit global average temperature increase to below 1.5 °C above pre-industrial levels in order to transition towards deep-decarbonization' (para 19a). The commitment at Paris fell short of the expectations of the PIDF. The 2 °C will not halt the threats to low-lying islands. In other words, this target will not increase climate security in the Pacific. Secondly, the PIDF called for 'loss and damage to be anchored as a standalone element that is separate and distinct from adaptation in the 2015 Paris Climate Change Agreement' (para 19 d). Article 8 of the Paris Agreement recognizes the principle of loss and damage and provides enhanced status to the Warsaw International Mechanism for Loss and Damage associated with the impact of climate change. However, as Hoad (2016: 318) notes the acceptance of loss and damage is not unequivocal since the preamble to the Paris Agreement precludes states from seeking liability or compensation for loss and damage. The third important 'demand' was the call for 'increased support for adaptation measures that address all vulnerable sectors including health, water and sanitation, energy, agriculture, forestry and fisheries' (para 19h). The Paris Agreement includes a commitment to meet the Cancun target of $100 billion per year in climate finance and extended this to 2025. Moreover, states were urged to place a higher priority on adaptation funding.

It is too early to develop firm conclusions on the impact of the Paris Agreement on Pacific Island nations. On one hand one could conclude that the agreement 'appears to give small island states easier access to funding and to newly invigorated clean development mechanisms, and acknowledges the need for global funding to support the adaptation and mitigation plans outlined in their INDCs' (Hoad 2016: 319). On the other hand, it has been argued that 'the Paris Agreement offered new signs of a willingness to move the world in the right direction but there's an urgent need to turn this into concrete action' (Maclellan and Meads 2016: 66). In other words, the commitments made at Paris may be necessary but they are not sufficient to produce enhanced climate security in the Pacific.

Foundations for Future Research

Climate change builds upon regional solidarities in the Pacific while at the same time extending these relationships, imputing them with an unprecedented degree of significance and responsibility in helping to save the region from extinction. The origins of regional governance were explored in Chap. 2, and we have identified the ways decolonization, dependency, and a shared sense of culture, the Pacific Way, have shaped the development of regional governance. Climate change has expanded the capacity and responsibility of regional organizations in the Pacific and, as we argue in the chapter, the foundations of regional identity based on a shared fate and a shared way of contending with problems. However, there are certainly limitations. Regional organizations lack capacity and the influx of funding, expertise, and advice from outside the region to address climate change has exacerbated this and created new dependencies. This accentuates the need for a regional rather than national response to climate change and exposes the dual vulnerability of PICTs; they are extremely vulnerable to climate change and extremely vulnerable to dependencies on donors.

During the coming decade it appears that regional organizations will continue to be at the forefront of climate governance in the Pacific. The regional architecture is supported and necessary, despite the limitations. However, there are signs of discontent in the regional architecture, particularly when it comes to checking the power of Australia and New Zealand, and in intra-regional dynamics such as the expulsion of Fiji for five years and the formation of the Melanesian Spearhead Group and other sub-regional networks. However, none of these appears capable of unravelling the architecture of regional governance, especially when it comes to climate change. At the same time regional organizations have become more inclusive, with the PIF expanding membership in recent years. Further research on the resoluteness of regional institutions is warranted, however, especially vis-à-vis mounting frustrations following the Paris Agreement and the challenges of climate finance to the region.

The Pacific is constructed as the frontline of climate change. This construction draws on the ways climate change narratives have emerged in the Pacific, and globally, and the ways these narratives have gained hegemony in considerations of development, the environment, migration, and the very survival of PICTs. As discussed in Chap. 3, as climate change comes to almost wholly define the Pacific as a region, other pressing issues are either ignored or absorbed into the ways climate change is addressed, governed, and financed. Critical voices have emerged from within the region (and outside it) challenging the hegemony of climate change and the impacts of this hegemony on governments and communities in the region. It is crucial to stress that these are not the voices of climate change deniers, rather these voices offer critical perspectives on the ways the hegemony of climate change has impacted the capacity of locals to contend with it and to contend with other pressing issues that may not fit neatly into the climate change narrative. While we have discussed this in some detail, this is certainly an avenue for further research at the local level in the Pacific. At the regional level, our main focus, critical voices are few. While respondents we interviewed often gave critical viewpoints about the ways climate change is politicized and governed in the Pacific from time to time, critical views were most commonly expressed 'off the record' or focused on the lack of finance, the complex institutional arrangements, or the inflexibility of donors. Rarely was there open questioning of dominant climate change narratives or the need for a regional response. This brings us to a related argument developed in Chap. 3; the construction of the Pacific as the frontline of climate change has been utilized effectively by PICTs, especially in international climate change meetings and negotiations, and as part of AOSIS. There is simply too much to lose for PICT governments and regional organizations by questioning the impacts of climate change narratives openly.

Constructing climate change as a security threat focuses on one of the climate change narratives and the ways it has brought the Pacific into discussions about regional stability, failed states, and refugee crises. These discussions also take place in different arenas, such as the United Nations Security Council, and among different actors, such as security experts rather than environment and development experts. Using the conceptual vocabulary of the Copenhagen School and the broader literature on environmental security, we used Chap. 4 to discuss the framing of climate change as a security threat in the Pacific. We identified two competing discourses of climate security, climate change as a threat to peace and climate change as exacerbating vulnerabilities within and between PICTs. We argued that governments in the Pacific, and climate activists to a certain extent, draw on both of these competing discourses where necessary. Here too there are possibilities for further research, particularly as the climate security discourse continues to proliferate and permeate in the region. The extent to which this is internalized within domestic institutions in PICTs deserves further research to explore the potential divergence and convergence in the ways climate security is invoked away from the regional consensus and in the context of competing domestic interests. There is also an opportunity to connect the securitization of climate change with access to aid and finance external to the global climate finance architecture discussed in Chap. 5. Does securitizing climate change directly relate to the ways it is financed? This question needs detailed attention.

The regional architecture remains the most favoured mechanism to deliver and distribute climate finance, the focus of Chap. 5. PICTs have had a role in shaping the global financial architecture, and the notion of the frontline, a notion we foregrounded at the start of the book, is crucial in giving PICTs voice in this process. Climate finance extends the notion of climate justice. PICTs contributed little to the human activities that have unleashed climate change and have a very limited capacity to addresses their impacts, thus the global and regional climate finance regime seeks to transfer resources to the region, through regional organizations. While few would contend this principle, the process has been far more complex and convoluted. As we discuss in the chapter, the failure to reach agreement over the objectives of climate finance, the required volume of finance, and the governance of climate finance demonstrate this complexity. Despite agreement on the need for climate finance in the region, the current state of climate finance negatively impacts the ability of PICTs and other small island states to secure funds to combat climate change. This part of our study changes rapidly. Further research to capture the dynamics of climate finance, including the changing demand, the successes, and the shortcomings will seemingly be necessary. Close monitoring of the ways finance is discussed at the various COP meetings and the possibility of increased (or reduced) bilateral finance will also be valuable. So too will the capacity for regional organizations in the Pacific and individual PICTs to utilize the finance they receive. This was a common problem identified by respondents in our research. Wide recognition of the need for more funds is accompanied by anxiety that these funds require human and material resources that are often not present at the local level or unevenly distributed. This is certainly an avenue to follow in future research on climate finance in the Pacific.

Reference

Dimitrov, R. 2016. The Paris Agreement on Climate Change: Behind Closed Doors. *Global Environmental Politics* 16 (3): 1–11.

Hoad, D. 2016. The 2015 Paris Climate Agreement: Outcomes and Their Impacts on Small Island states. *Island Studies Journal* 11 (1): 315–320.

Maclellan, N., and S. Meads. 2016. *After Paris: Climate Finance in the Pacific Islands: Strengthening Collaboration, Accelerating Access and Prioritising Adaptation for Vulnerable Communities.* Auckland: Oxfam New Zealand/Oxfam Australia.

Spash, C.L. 2016. This Changes Nothing: The Paris Agreement to Ignore Reality. *Globalizations* 13 (6): 928–933.

UNFCCC. 2015. Historic Paris Agreement on Climate Change: 195 Nations Set Path to Keep Temperature Rise Well Below 2 Degrees Celsius [Online]. Available at: http://newsroom.unfccc.int/unfccc-newsroom/finale-cop21/. Accessed 12 Aug 2016.

References

Adger, W.N., S. Huq, K. Brown, D. Conway, and M. Hulme. 2003. Adaptation to Climate Change in the Developing World. *Progress in Development Studies* 3 (3): 179–195.

Aisi, R. 2015. *Statement for the UN Security Council, Open Arria-Formula Meeting on the Role of Climate Change as a Threat Multiplier for Global Security.* Available at: http://www.spainun.org/wp-content/uploads/2015/07/Papua-New-Guinea_CC_201506.pdf. Accessed 20 June 2017.

Amerasinghe, N., J. Thwaites, G. Larsen, and A. Ballesteros. 2017. *Future of the Funds: Exploring the Architecture of Multilateral Climate Finance.* Washington, DC: WRI.

AOSIS. 2015. *AOSIS Opening Statement for 21st Conference of Parties to the UNFCCC.* Available at: http://aosis.org/wp-content/uploads/2015/12/FINAL-AOSIS-COP-Statement-Paris-.pdf. Accessed 28 July 2016.

AOSIS. n.d. About AOSIS. Available at: http://aosis.org/about-aosis/. Accessed 19 Feb 2014.

Ashe, J.W., R. Van Lierop, and A. Cherian. 1999. The Role of the Alliance of Small Island States (AOSIS) in the Negotiation of the United Nations Framework Convention on Climate Change (UNFCCC). *Natural Resources Forum* 23: 209–220.

Asian Development Bank (ADB). 2005. *Toward a New Pacific Regionalism.* Available at: http://www.adb.org/publications/toward-new-pacific-regionalism. Accessed 30 Aug 2013.

Asia Development Bank (ADB). 2007a. *Japan and Australia Extend $88M in Financial Assistance to Samoa Power Sector.* Asia Development Bank. http://www.adb.org/media/printer.asp?articleID=12273. Accessed 15 Aug 2016.

M. Williams, D. McDuie-Ra, *Combatting Climate Change in the Pacific*, https://doi.org/10.1007/978-3-319-69647-8

———. 2007b. *ADB, Japan and Australia Extend $88M in Financial Assistance to Samoa Power Sector*. Available at: http://www.adb.org/media/printer.asp?articleID=12273. Accessed 30 Aug 2013.

———. n.d. *Project Information Document 38183: Power Sector Project* (Formerly Supporting Energy Sector Reforms). Asia Development Bank. http://www.adb.org/projects/project.asp?id=38183. Accessed 15 Aug 2016.

Atkinson, J. 2007. Vanuatu in Australia-China-Taiwan Relations. *Australian Journal of International Affairs* 61 (3): 351–366.

———. 2010. China–Taiwan Diplomatic Competition and the Pacific Islands. *The Pacific Review 23* (4): 407–427.

Atterridge, A., C.K. Siebert, R.J.T. Klein, C. Butler, and P. Tellar. 2009. *Bilateral Finance Institutions and Climate Change: A Mapping of Climate Portfolios*. Stockholm: Stockholm Environment Institute.

AusAid. 2007. *Aid and the Environment—Building Resilience, Sustaining Growth: An Environment Strategy for Australian Aid*. Canberra: Commonwealth of Australia.

Australian Bureau of Meteorology and CSIRO. 2014. *Climate Change in the Pacific: Scientific Assessment and New Research*. Vol. 1, *Regional Overview*. Canberra: Australian Bureau of Meteorology.

Australian Government. 2009. *Engaging Our Pacific Neighbours on Climate Change: Australia's Approach*. Canberra: Commonwealth of Australia.

Australian Government, Department of Defence. 2003. *Australia's National Security: A Defence Update 2003*. Canberra: Commonwealth of Australia.

———. 2005. *Australia's National Security: A Defence Update 2005*. Canberra: Commonwealth of Australia.

———. 2009. *Defending Australia in the Asia Pacific Century: Force 30. Defence White Paper 2009*. Canberra: Commonwealth of Australia.

Baker, N. 2015. New Zealand and Australia in Pacific Regionalism. In *The New Pacific Diplomacy*, ed. G. Fry and S. Tarte, 137–148. Canberra: ANU E-Press.

Balzacq, T. 2016. The Three Faces of Securitization: Political Agency, Audience and Context. *European Journal of International Relations* 11 (2): 171–201.

Barbara, J., and H. McMahon. 2016. *Pacific Regional Youth Employment Scan*. Suva: Pacific Leadership Program.

Barnett, J. 2001. Adapting to Climate Change in Pacific Island Countries: The Problem of Uncertainty. *World Development 29* (6): 977–993.

———. 2003. Security and Climate Change. *Global Environmental Change 13* (1): 7–17.

———. 2005. Titanic States? Impacts and Responses to Climate Change in the Pacific Islands. *Journal of International Affairs* 59 (1): 203–219.

Barnett, J., and W.N. Adger. 2003. Climate Dangers and Atoll Countries. *Climatic Change* 61 (3): 321–337.

———. 2007. Climate Change, Human Security and Violent Conflict. *Political Geography* 26 (6): 639–655.

Barnett, J., and J. Campbell. 2010. *Climate Change and Small Island States: Power, Knowledge, and the South Pacific*. London/Washington, DC: Earthscan.
Barnett, J. 2010. Dangerous Climate Change in the Pacific Islands: Food Production and Food Security. *Regional Environmental Change* 11 (1): 229–237.
Bell, D. 2013. Climate Change and Human Rights. *Wiley Interdisciplinary Reviews: Climate Change* 4 (3): 159–170.
Bellamy, P. 2008. The 2006 Fiji Coup and Impact on Human Security. *Journal of Human Security* 4 (2): 4–18.
Benwell, R. 2011. The Canaries in the Coalmine: Small States as Climate Change Champions. *The Round Table* 100 (413): 199–211.
Bertram, G. 1999. The MIRAB Model Twelve Years On. *The Contemporary Pacific* 11 (1): 105–138.
———. 2006. Introduction: The MIRAB Model in the Twenty-First Century. *Asia Pacific Viewpoint* 47 (1): 1–13.
Bertram, G., and R.F. Watters. 1986. The MIRAB Process: Earlier Analyses in Context. *Pacific Viewpoint* 27 (1): 47–59.
Betzold, C. 2010. 'Borrowing' Power to Influence International Negotiations: AOSIS in the Climate Change Regime, 1990–1997. *Politics* 30 (3): 131–148.
Betzold, C., P. Castro, and F. Weiler. 2012. AOSIS in the UNFCCC Negotiations: From Unity to Fragmentation? *Climate Policy* 12 (5): 591–613.
Booth, K. 1991. Security in Anarchy: Utopian Realism in Theory and Practice. *International Affairs* 67 (3): 527–545.
Boyd, E., N. Grist, S. Juhola, and V. Nelson. 2009. Exploring Development Futures in a Changing Climate: Frontiers for Development Policy and Practice. *Development Policy Review* 47 (6): 659–674.
Brant, P. 2013. Chinese Aid in the South Pacific: Linked to Resources? *Asian Studies Review* 37 (2): 158–177.
Brindis, D. 2007. What Next for the Alliance of Small Island States in the Climate Change Arena? *Sustainable Development Law and Policy* 7 (2): 45–85.
Brown, M.E., and C. Funk. 2008. *Food Security Under Climate Change*. National Aeronautics and Space Administration. University of Nebraska–Lincoln Paper 131. Available at: http://digitalcommons.unl.edu. Accessed 12 July 2015.
Brown, O., A. Hammill, and R. McLeman. 2007. Climate Change as the 'New' Security Threat: Implications for Africa. *International Affairs* 83 (6): 1141–1154.
Browne, C., and A. Mineshima. 2007. Remittances in the Pacific Region. IMF Working Paper (WP/07/35).
Bryant-Tokalau, J.J. 1995. The Myth Exploded: Urban Poverty in the Pacific. *Environment and Urbanization* 7 (2): 109–130.
Buzan, B. 1991. New Patterns of Global Security in the Twenty-First Century. *International Affairs 67* (3): 431–451. Waever.

Buzan, B., O. Weaver, and J. de Wilde. 1998. *Security: A New Framework for Analysis*. Boulder/London: Lynne Rienner.

Cannon, T., and D. Müller-Mahn. 2010. Vulnerability, Resilience and Development Discourses in Context of Climate Change. *Natural Hazards* 55 (3): 621–635.

Carvani, A., C. Watson, and L. Schalatek. 2016. *Climate Finance Thematic Briefing: Adaptation Finance*. London/Washington, DC: ODI and Heinrich Böll Stiftung.

Chand, S. 2010. Shaping New Regionalism in the Pacific Islands: Back to the Future? (No. 61). ADB Working Paper Series on Regional Economic Integration.

Chasek, P.S. 2005. Margins of Power: Coalition Building and Coalition Maintenance of the South Pacific Island States and the Alliance of Small Island States. *Review of European, Comparative and International Environmental Law* 14 (2): 125–137.

Connell, J. 1993. Climatic Change: A New Security Challenge for the Atoll States of the South Pacific. *Journal of Commonwealth and Comparative Politics 31* (2): 173–192.

———. 2003. Losing Ground? Tuvalu, the Greenhouse Effect and the Garbage Can. *Asia Pacific Viewpoint 44* (2): 89–107.

———. 2010. From Blackbirds to Guestworkers in the South Pacific. *Plus ça Change…? The Economic and Labour Relations Review: ELRR* 20 (2): 111–121.

———. 2011. Elephants in the Pacific? Pacific Urbanisation and Its Discontents. *Asia Pacific Viewpoint 52* (2): 121–135.

———. 2015. Vulnerable Islands: Climate Change, Tectonic Change, and Changing Livelihoods in the Western Pacific. *The Contemporary Pacific* 27 (1): 1–36.

Connell, J., and J. Lea. 2002. *Urbanisation in the Island Pacific: Towards Sustainable Development*. London/New York: Routledge.

Corbett, J. 2015. "Everybody Knows Everybody": Practising Politics in the Pacific Islands. *Democratization* 22 (1): 51–72.

Crocombe, R. 1976. *The Pacific Way: An Emerging Identity*. Suva: Lotu Pasifika Productions.

Dahl, A.L., and I.L. Baumgart. 1983. *The State of the Environment in the South Pacific*. Nairobi: UNEP.

Dalby, S. 2009. *Security and Environmental Change*. Cambridge: Polity.

Davis, W.J. 1996. The Alliance of Small Island States (AOSIS): The International Conscience. *Asia-Pacific Magazine* 2 (May): 17–22. Available at: http://coombs.anu.edu.au/SpecialProj/APM/TXT/davis-j-02-96.html. Accessed 22 Feb 2014.

Deitelhoff, N., and L. Wallbott. 2012. Beyond Soft Balancing: Small States and Coalition-Building in the ICC and Climate Negotiations. *Cambridge Review of International Affairs 25* (3): 345–366.

Denton, F. 2002. Climate Change Vulnerability, Impacts, and Adaptation: Why Does Gender Matter? *Gender and Development* 10 (2): 10–20.
Detraz, N. 2009. Environmental Security and Gender: Necessary Shifts in an Evolving Debate. *Security Studies* 18 (2): 345–369.
———. 2011. Threats or Vulnerabilities? Assessing the Link Between Climate Change and Security. *Global Environmental Politics* 11 (3): 104–120.
Detraz, N., and M.M. Betsill. 2009. Climate Change and Environmental Security: For Whom the Discourse Shifts. *International Studies Perspectives* 10 (3): 303–320.
Dimitrov, R. 2016. The Paris Agreement on Climate Change: Behind Closed Doors. *Global Environmental Politics* 16 (3): 1–11.
Dinnen, S. 2002. Winners and Losers: Politics and Disorder in the Solomon Islands 2000–2002. *The Journal of Pacific History* 37 (2): 285–298.
Dupont, A., and G. Pearman. 2006. *Heating Up the Planet: Climate Change and Security*. Vol. 12. Sydney: Lowy Institute for International Policy.
Durand, A., L. Schalatek, and C. Watson. 2015. *Climate Finance Briefing: Small Island Developing States*. London/Washington, DC: ODI and Heinrich Böll Stiftung.
Dyer, M. 2017. Eating Money: Narratives of Equality on Customary Land in the Context of Natural Resource Extraction in the Solomon Islands. *The Australian Journal of Anthropology* 28 (1): 88–103.
Edwards, M.J. 1996. Climate Change, Worst-Case Analysis and Ecocolonialism in the Southwest Pacific. *Global Change, Peace and Security* 8 (1): 63–80.
———. 1999. Security Implications of a Worst-Case Scenario of Climate Change in the South-West Pacific. *Australian Geographer* 30 (3): 311–330.
Ehlers, W. 2011. *Institutional Structure of the GEF*. GEF Expanded Constituency Workshop, Global Environment Facility, March 1–3, 2011, Belize City, Belize. http://www.thegef.org/gef/node/4233. Accessed 15 Aug 2016.
Elliott, L. 2007. Environment and Security: What's the Connection? *Australian Defence Force Journal* 174: 39–52.
———. 2010. Climate Migration and Climate Migrants: What Threat, Whose Security? In *Climate Change and Displacement: Multidisciplinary Perspectives*, ed. Jane McAdam, 175–190. Oxford/Portland: Hart Publishing.
FAO. n.d. Climate Change and Food Security. Available at: http://www.fao.org/climatechange. Accessed 23 Sep 2016.
Farbotko, C. 2005. Tuvalu and Climate Change: Constructions of Environmental Displacement in the Sydney Morning Herald. *Geografiska Annaler: Series B, Human Geography* 87 (4): 279–293.
———. 2010. Wishful Sinking: Disappearing Islands, Climate Refugees and Cosmopolitan Experimentation. *Asia Pacific Viewpoint* 51 (1): 47–60.
Federal Republic of Germany. 2016. *Climate Change Protection Through Forest Conservation in Pacific Island States*. Berlin: Federal Ministry for the Environment, Nature Conservation, Building and Nuclear Safety.

FFA. 2017. Economic and Development Indicators and Statistics: Tuna Fisheries of the West and Central Pacific Ocean 2016. Available at: https://www.ffa.int/system/files/FFA%20Economic%20and%20Development%20Indicators%20and%20Statistics%202016.pdf. Accessed 2 Aug 2017.

Fisher, P.B. 2011. Climate Change and Human Security in Tuvalu. *Global Change, Peace and Security 23* (3): 293–313.

Food and Agriculture Organization (FAO). 2008a. *Climate Change and Food Security in Pacific Island Countries.* Rome: Food and Agricultural Organization.

———. 2008b. *Climate Change and Food Security: A Framework Document.* Available at: http://www.fao.org/forestry/15538-079b31d45081fe9c3db-c6ff34de4807e4.pdf. Accessed 23 Sept 2016.

Frazer, I., and J. Bryant-Tokalau. 2006. Introduction: The Uncertain Future of Pacific Regionalism. In *Redefining the Pacific?: Regionalism Past, Present and Future*, ed. J. Bryant-Tokalau and I. Frazer, 1–24. Aldershot: Ashgate.

FRDP. 2016. *Framework for Resilient Development in the Pacific: An Integrated Approach to Climate Change and Disaster Risk Management (FRDP) 2017–2030.* Available at: http://www.forumsec.org/resources/uploads/embeds/file/Annex%201%20-%20Framework%20for%20Resilient%20Development%20in%20the%20Pacific.pdf. Accessed 12 July 2017.

GEF. 2017. Report of the Global Environment Facility to the Twenty-third Session of the Conference of the Parties to the United Nations Framework Convention on Climate Change. Available at: https://www.thegef.org/sites/default/files/documents/Final%20COP%2023%20Report%20August%203.pdf. Accessed 10 Aug 2017.

Grasso, M. 2006. An Ethics-Based Climate Agreement for the South Pacific Region. *International Environmental Agreements: Politics, Law and Economics* 6 (3): 249–270.

Green Climate Fund. 2017. *GCF in Numbers.* http://www.greenclimate.fund/documents/20182/24871/GCF_in_Numbers.pdf/226fc825-3c56-4d71-9a4c-60fd83e5fb03. Accessed 12 July 2017.

Grist, N. 2008. Positioning Climate Change in Sustainable Development Discourse. *Journal of International Development* 20 (6): 783–803.

Grydehøj, A., and I. Kelman. 2017. The Eco-Island Trap: Climate Change Mitigation and Conspicuous Sustainability. *Area* 49 (1): 106–113.

Haines, A., and J.A. Patz. 2004. Health Effects of Climate Change. *Journal of the American Medical Association* 291 (1): 99–103.

Haines, A., R.S. Kovats, D. Campbell-Lendrum, and C. Corvalan. 2006. Climate Change and Human Health: Impacts, Vulnerability, and Mitigation. *Lancet* 367 (9528): 2101–2109.

Haites, E. 2014. *Aligning Climate Finance and Development for Asia and the Pacific: Potential and Prospects Manila.* Asian Development Bank. https://www.adb.org/sites/default/files/publication/152437/sdwp-033.pdf. Accessed 4 Aug 2016.

Harrison, D.H., and B.C. Prasad. 2013. The Contribution of Tourism to the Development of Fiji and Other Pacific Island Countries. In *Handbook of Tourism Economics, Analysis, Applications, Case Studies*, ed. Clement A. Tisdell, 741–761. Singapore: World Scientific Publishing.

Hayward, B. 2008. Let's Talk About the Weather: De-Centering Democratic Debate About Climate Change. *Hypatia: A Journal of Feminist Philosophy* 23 (3): 79–98.

Hendrix, C.S., and S.M. Glaser. 2007. Trends and Triggers: Climate, Climate Change and Civil Conflict in Sub-Saharan Africa. *Political Geography 26* (6): 695–715.

Herr, R., and A. Bergin. 2011. *Our Near Abroad: Australia and Pacific Islands Regionalism*. Barton: Australian Strategic Policy Institute.

Hoad, D. 2016. The 2015 Paris Climate Agreement: Outcomes and Their Impacts on Small Island States. *Island Studies Journal* 11 (1): 315–320.

Homer-Dixon, T.F. 1991. On the Threshold: Environmental Changes as Causes of Acute Conflict. *International Security* 16 (2): 76–116.

Hsiang, S.M., K.C. Meng, and M.A. Cane. 2011. Civil Conflicts Are Associated with the Global Climate. *Nature* 476 (7361): 438–441.

Hviding, E. 2003. Contested Rainforests, NGOs, and Projects of Desire in Solomon Islands. *International Social Science Journal* 55 (178): 539–554.

INCFCCC. 1991. *Statement on Commitments Submitted by the Delegation of Vanuatu on Behalf of the States Members of the AOSIS, in Intergovernmental Negotiating Committee for a Framework Convention on Climate Change*. A/AC.237/WG.1/L.9.

———. 1994. *Matters Relating to Arrangements for the Financial Mechanism Implementation of Article 11, Paras. 1–4, in Intergovernmental Negotiating Committee for a Framework Convention on Climate Change*. A/AC.237/MISC.40.

International Climate Change Adaptation Initiative (ICCAI). 2011. *Climate Change in the Pacific: Scientific Assessment and New Research. Volume 1: Regional Overview*. Available at: http://www.cawcr.gov.au/projects/PCCSP/Nov/Vol1_CoversForewordContents.pdf. Accessed 29 May 2013.

International Institute for Sustainable Development (IISD). 1994. *Summary of the UN Global Conference on The Sustainable Development of Small Island Developing States: 25 April–6 May 1994*. Available at: http://www.iisd.ca/vol08/0828000e.html. Accessed 19 Feb 2014.

IPCC. 2013. *Working Group I Contribution to the Fifth Assessment Report of the Intergovernmental Panel on Climate Change 2013: Summary for Policymakers*.

Kaplan, R.D. 1994. The Coming Anarchy. *The Atlantic Monthly, February* 273: 44–76.

Kelman, I. 2014. No Change from Climate Change: Vulnerability and Small Island Developing States. *The Geographical Journal* 180 (2): 120–129.

Kelman, I., R. Stojanov, S. Khan, O.A. Gila, B. Duží, and D. Vikhrov. 2015. Viewpoint Paper. Islander Mobilities: Any Change from Climate Change? *International Journal of Global Warming* 8 (4): 584–602.

Kempf, W. 2009. A Sea of Environmental Refugees? Oceania in an Age of Climate Change. In *Form, Macht, Differenz: Motive und Felder Ethnologischen Forschens*, ed. E. Hermann, K. Klenke, and M. Dickhardt, 191–205. Göttingen: Universitätsverlag Göttingen.

Kendall, R. 2012. Climate Change as a Security Threat to the Pacific Islands. *New Zealand Journal of Environmental Law* 16: 83–116.

Koser, K. 2012. *Environmental Change and Migration: Implications for Australia*. Sydney: Lowy Institute for International Policy. http://www.lowyinstitute.org/publications/environmental-change-and-migration-implications-australia. Accessed 12 July 2017.

Komai, M. 2013. *Reconfiguring Regionalism in the Pacific*. Port Vila: Pacific Institute of Public Policy.

Lane, R., and R. McNaught. 2009. Building Gendered Adaptation to Climate Change in the Pacific. *Gender and Development* 17 (1): 67–80.

Lawson, S. 2010. 'The Pacific Way' as Postcolonial Discourse: Towards a Reassessment. *The Journal of Pacific History* 45 (3): 297–314.

Lazrus, H. 2012. Sea Change: Island Communities and Climate Change. *Annual Review of Anthropology* 41: 285–301.

———. 2016. Shifting Tides: Climate Change, Migration, and Agency in Tuvalu. In *Anthropology and Climate Change: From Actions to Transformations*, ed. S. Crate and M. Nuttall, 2nd ed., 261–270. London: Routledge.

Locke, J.T. 2009. Climate Change-Induced Migration in the Pacific Region: Sudden Crisis and Long-Term Developments. *The Geographical Journal* 175 (3): 171–180.

Lomborg, Bjorn. 2016. About Those Non-Disappearing Pacific Islands. *The Wall Street Journal*, October 13.

Lovell, S.A. 2011. Health Governance and the Impact of Climate Change on Pacific Small-Island Developing States. *IHDP Update* 1: 50–55.

Maclellan, N. 2011. *Improving Access to Climate Financing for the Pacific Islands*. Sydney: Lowy Institute.

Maclellan, N., and S. Meads. 2016. *After Paris: Climate Finance in the Pacific Islands: Strengthening Collaboration, Accelerating Access and Prioritising Adaptation for Vulnerable Communities*. Auckland: Oxfam New Zealand/Oxfam Australia.

Makun, K.K. 2017. Imports, Remittances, Direct Foreign Investment and Economic Growth in Republic of Fiji Islands: An Empirical Analysis Using ARDL Approach? *Kasetsart Journal of Social Sciences* (Online First). https://doi.org/10.1016/j.kjss.2017.07.002.

Mara, R.K. 1997. *The Pacific Way: A Memoir*. Honolulu: University of Hawai'i Press.

Marawa, S. 2015. Negotiating the Melanesia Free Trade Area. In *The New Pacific Diplomacy*, ed. G. Fry and S. Tarte, 161–174. Canberra: ANU E-Press.

McCubbin, S., B. Smit, and T. Pearce. 2015. Where Does Climate Fit? Vulnerability to Climate Change in the Context of Multiple Stressors in Funafuti, Tuvalu. *Global Environmental Change* 30: 43–55.

McDonald, M. 2013. Discourses of Climate Security. *Political Geography* 33: 42–51.

McMichael, A., R. Woodruff, and S. Hales. 2006. Climate Change and Human Health: Present and Future Risks. *The Lancet* 367 (9513): 859–869.

Michaelowa, A., and K. Michaelowa. 2007a. Climate or Development: Is ODA Diverted from Its Original Purpose? *Climatic Change* 84 (1): 5–21.

———. 2007b. Does Climate Policy Promote Development? *Climatic Change* 84 (1): 1–4.

Mishra, S. 2005. Pacific Way. In *A Historical Companion to Postcolonial Though in English*, ed. Prem Poddar and David Johnson, 364–368. New York: Columbia University Press.

Moore, E.J., and J.W. Smith. 1995. Climatic Change and Migration from Oceania: Implications for Australia, New Zealand and the United States of America. *Population and Environment* 17 (2): 105–122.

Mortreux, C., and J. Barnett. 2009. Climate Change, Migration and Adaptation in Funafuti, Tuvalu. *Global Environmental Change* 19 (1): 105–112.

Nakhooda, et al. 2013. *Mobilising International Climate Finance: Lessons from the Fast-Start Finance Period*. Overseas Development Institute: London.

Nakhooda, S., C. Watson, and L. Schalatek. 2016. *The Global Climate Finance Architecture*. London/Washington, DC: ODI and Heinrich Böll Stiftung.

Nunn, P.D. 2013. The End of the Pacific? Effects of Sea Level Rise on Pacific Island Livelihoods. *Singapore Journal of Tropical Geography* 34 (2): 143–171.

O'Brien, R., and M. Williams. 2010. *Global Political Economy: Evolution and Dynamics*. 3rd ed. Houndmills: Palgrave Macmillan.

OECD. 2015. *Climate Finance in 2013–14 and the USD 100 Billion Goal: A Report by the OECD in Collaboration with Climate Policy Initiative*. Paris: OECD Publishing. https://doi.org/10.1787/9789264249424-en.

———. n.d. *Poverty and Climate Change: Reducing the Vulnerability of the Poor Through Adaptation*. Available at: http://www.oecd.org/dataoecd/60/27/2502872.pdf. Accessed 12 Mar 2014.

Oxfam. 2009. *The Future Is Here: Climate Change in the Pacific*. Carlton/Newton: Oxfam Australia/Oxfam New Zealand.

Pacific Community. 2016. *Pacific Community: Financial Statements for 2015*. Available at: http://www.spc.int/wp-content/uploads/2017/02/Annual-Report-EN-2015-V2.pdf. Accessed 31 Aug 2017.

PIDF. 2015. *Suva Declaration on Climate Change*. http://pacificidf.org/wp-content/uploads/2013/06/Suva-Climate-Declaration-final_USB.pdf. Accessed 2 July 2017.

Pacific Islands Forum (PIF). 1994. Forum Communiqué. 25th Pacific Islands Forum, 31 July–2 August, Brisbane, Australia. Available at: http://forum.forumsec.org/resources/uploads/attachments/documents/1994%20Communique Brisbane%2031Jul-2%20Aug.pdf. Accessed 18 June 2011.

Pacific Islands Forum (PIF). 1988. Forum Communiqué. 19th Pacific Islands Forum, 20–21 September, Nuku'alofa, Tonga. Available at: http://forum.forumsec.org/resources/uploads/attachments/documents/1988%20Communique Tonga%2020-2%20Sept.pdf. Accessed 21 April 2010.

Pacific Islands Forum (PIF). 2007. "Forum Leaders Communiqué" 38th Pacific Islands Forum, 16–17 October, Nuku'alofa, Tonga. Available at: http://www.forumsec.org/resources/uploads/attachments/documents/2007%20Forum%20Communique,%20Vava%27u,%20Tonga,%2016-17%20Oct.pdf. Accessed 21 Apr 2010.

———. 2008. "Forum Communique" 39th Pacific Islands Forum, 19–20 August, Alofi, Niue. Available at: http://forum.forumsec.org/resources/uploads/attachments/documents/2008%20Forum%20Communique,%20Alofi,%20Niue,%2019-20%20Aug.pdf. Accessed 4 Dec 2011.

———. 2009. "Forum Communique" 40th Pacific Islands Forum, 5–6 August, Cairns, Australia. Available at: http://forum.forumsec.org/resources/uploads/attachments/documents/2009%20Forum%20Communique,%20Cairns,%20Australia%205-6%20Aug.pdf. Accessed 4 Dec 2011.

———. 2010. *Pacific Plan Annual Progress Report.* Available at: http://www.forumsec.org/resources/uploads/attachments/documents/Pacific%20Plan%202010%20Annual%20Progress%20Report_Eng.pdf. Accessed 7 Aug 2014.

Pacific Islands Forum Secretariat (PIFS). 2003/2004. *Annual Report: Excelling Together for the People of the Pacific.* Available at: http://www.forumsec.org/resources/uploads/attachments/documents/2003-2004_PIFS_Annual_Report.pdf. Accessed 2 June 2011.

———. 2005a. *The Pacific Plan for Strengthening Regional Cooperation and Integration.* Suva: Fiji Pacific Islands Forum Secretariat.

———. 2005b. *The Pacific Plan.* Available at: http://www.sopac.org/sopac/docs/RIF/03_A_Pacific_Plan-2005.pdf. Accessed 22 Aug 2012.

———. 2010. *Pacific Plan Annual Progress Report.* Available at: http://www.forumsec.org/resources/uploads/attachments/documents/Pacific%20Plan%202010%20Annual%20Progress%20Report_Eng.pdf. Accessed 22 Sept 2013.

———. 2013. *Pacific Plan Review 2013: Report to the Leaders.* Available at: http://www.cid.org.nz/assets/Key-issues/Pacific-development/Pacific-Plan-Review-2013-Volume-2.pdf. Accessed 3 Aug 2014.

———. n.d. *Mission and Vision.* Available at: http://www.forumsec.org/pages.cfm/about-us/mission-goals-roles/?printerfriendly=true. Accessed 31 Aug 2017.

Parsons, R.J. 2011. Strengthening Sovereignty: Security and Sustainability in an Era of Climate Change. *Sustainability* 3 (9): 1416–1451.

Patel, S., C. Watson, and L. Schalatek. 2016. *Climate Finance Thematic Briefing: Mitigation Finance*. London/Washington, DC: ODI and Heinrich Böll Stiftung.

Pernetta, J.C., and P.J. Hughes. 1990. *Implications of Expected Climate Changes in the South Pacific Region: An Overview*. Nairobi: UNEP.

PIF. 1988. "Forum Communique" 19th Pacific Islands Forum, 20–21 September, Nuku'alofa, Tonga. Available at: http://forum.forumsec.org/resources/uploads/attachments/documents/1988%20Communique Tonga%2020-2%20 Sept.pdf. Accessed 21 April 2010.

———. 1994. "Forum Communique" 25th Pacific Islands Forum, 31 July–2 August, Brisbane, Australia. Available at: http://forum.forumsec.org/resources/uploads/attachments/documents/1994%20Communique Brisbane%2031Jul-2%20Aug.pdf. Accessed 18 June 2011.

Podesta, J., and P. Ogden. 2008. The Security Implications of Climate Change. *Washington Quarterly* 31 (1): 115–138.

Polycarp, C. et al. 2012. *Developed Country Fast-Start Climate Finance Pledges: A Summary of Self-Reported Information*. World Resources Institute. http://www.wri.org/sites/default/files/pdf/climate_finance_pledges_2012-11-26.pdf.

Raleigh, C. 2010. Political Marginalization, Climate Change, and Conflict in African Sahel States. *International Studies Review* 12 (1): 69–86.

Reuveny, R. 2007. Climate Change-Induced Migration and Violent Conflict. *Political Geography* 26 (6): 656–673.

Rolfe, J. 2001. Peacekeeping the Pacific Way in Bougainville. *International Peacekeeping* 8 (4): 38–55.

Rosen, F., and P. Olsson. 2013. Institutional Entrepreneurs, Global Networks, and the Emergence of International Institutions for Ecosystem-Based Management: The Coral Triangle Initiative. *Marine Policy* 38: 195–204.

Roy, P., and J. Connell. 1991. Climatic Change and the Future of Atoll States. *Journal of Coastal Research* 7 (4): 1057–1075.

Rudiak-Gould, P. 2016. Climate Change Beyond the "Environmental": The Marshallese Case. In *Anthropology and Climate Change: From Actions to Transformations*, ed. S. Crate and M. Nuttall, 2nd ed., 220–227. London: Routledge.

Russell, L. 2011. *Poverty, Climate Change and Health in Pacific Island Countries: Issues to Consider in Discussion, Debate and Policy Development*. University of Sydney and Australian National University: Menzies Centre for Health and Public Policy. Available at: http://www.menzieshealthpolicy.edu.au/other_tops/pdfs_pubs/pacificislands2011.pdf. Accessed 23 Aug 2011.

Salehyan, I. 2008. From Climate Change to Conflict? No Consensus Yet. *Journal of Peace Research* 45 (3): 315–326.

Schalatek, L., and N. Bird. 2016. *The Principles and Criteria of Public Climate Finance—A Normative Framework*. London/Washington, DC: ODI and Heinrich Böll Stiftung.

Schalatek, L., and S. Nakhooda. 2016. *Gender and Climate Finance*. London/Washington, DC: ODI and Heinrich Böll Stiftung.

Schalatek, L., S. Nakhooda, and C. Watson. 2016. *The Green Climate Fund*. London/Washington, DC: ODI and Heinrich Böll Stiftung.

Scott, S.V. 2015. Implications of Climate Change for the UN Security Council: Mapping the Range of Potential Policy Responses. *International Affairs* 91 (6): 1317–1333.

SPC. 2011. Euro 11.4 M Climate Resilience Project Will Help Nine Pacific Small Island States 18 July, Secretariat of the Pacific Community. Available at: http://www.spc.int/en/component/content/article/740-114-m-climate-resilience project-will-help-nine-pacific-small-island-states.html. Accessed 7 Sep 2011.

SPC LRD. 2011. *SPC/GIZ Climate Protection Through Forest Conservation in the Pacific Islands, 25 July, Secretariat of the Pacific Community Land Resources Division*. Nabua: The Pacific Community Land Resources Division.

SPREP. 2014. *Secretariat of the Pacific Regional Environment Program Annual Report 2014*. Apia: Secretariat of the Pacific Regional Environment Program.

———. 2017. About PIGARREP. Available at: http://www.sprep.org/Pacific-Islands-Greenhouse-Gas-Abatement-through-Renewable-Energy-Project/about-piggarep. Accessed 31 Aug 2017.

Shibuya, E. 1997. Roaring Mice Against the Tide: The South Pacific Islands and Agenda-Building on Global Warming. *Pacific Affairs* 69 (4): 541–555.

Shibuya, E. 2009. The Problems and Potential of the Pacific Islands Forum. In *The Asia-Pacific: A Region in Transition*, ed. J. Rolf, 102–115. Honolulu: Asia-Pacific Centre for Security Studies.

Shie, T.R. 2007. Rising Chinese Influence in the South Pacific: Beijing's "Island Fever". *Asian Survey* 47 (2): 307–326.

Slade, S.G. 2009. *Statement by SG Slade at opening of 40th PIF*. Available at: http://lists.spc.int/pipermail/ppapd-fpocc/2009-August/000310.html. Accessed 25 Aug 2017.

Smith, P.J. 2007. Climate Change, Weak States and the 'War on Terrorism' in South and Southeast Asia. *Contemporary Southeast Asia* 29 (2): 264–285.

Spash, C.L. 2017. This Changes Nothing: The Paris Agreement to Ignore Reality. *Globalizations* 13 (6): 928–933.

Stern, N.H. 2007. *The Economics of Climate Change: The Stern Review*. Cambridge: Cambridge University Press.

Storey, D., and S. Hunter. 2010. Kiribati: An Environmental 'Perfect Storm'. *Australian Geographer* 41 (2): 167–181.

Strokirch, K.V. 2007. The Region in Review: International Issues and Events, 2005–2006. *The Contemporary Pacific* 19 (2): 552–577.

Tanner, T., and T. Mitchell. 2008. Entrenchment or Enhancement: Could Climate Change Adaptation Help to Reduce Chronic Poverty? *IDS Bulletin* 39 (4): 6–15.

Taplin, R. 1994. International Policy on the Greenhouse Effect and the Island South Pacific. *The Pacific Review* 7 (3): 271–281.

Tarte, S. 2014. Regionalism and Changing Regional Order in the Pacific Islands. *Asia and the Pacific Policy Studies* 1 (2): 312–324.

Tavola, K. 2015. Towards a New Regional Diplomacy Architecture. In *The New Pacific Diplomacy*, ed. G. Fry and S. Tarte, 27–38. Canberra: ANU E-Press.

Theisen, O.M. 2008. Blood and Soil? Resource Scarcity and Internal Armed Conflict Revisited. *Journal of Peace Research* 45 (6): 801–818.

Thomas, D.S., and C. Twyman. 2005. Equity and Justice in Climate Change Adaptation Amongst Natural-Resource-Dependent Societies. *Global Environmental Change* 15 (2): 115–124.

Thornton, A., M.T. Kerslake, and T. Binns. 2010. Alienation and Obligation: Religion and Social Change in Samoa. *Asia Pacific Viewpoint* 51 (1): 1–16.

Tisdell, C. 2008. Global Warming and the Future of Pacific Island Countries. *International Journal of Social Economics* 35 (12): 889–903.

UNDP. 1994. *Human Development Report*. Geneva: UNDP.

UNDP. 2007. *Human Development Report 2007/2008: Fighting Climate Change: Human Solidarity in a Divided World*. New York: United Nations Human Development Program.

UNEP. 2011. *Bilateral/Finance Institutions and Climate Change*. http://wedocs.unep.org/bitstream/handle/20.500.11822/7976/Mapping_report_final.pdf?sequence=3&isAllowed=y. Accessed 10 Mar 2016.

———. 2016. *The Adaptation Finance Gap Report 2016*. http://www.unep.org/adaptationgapreport/sites/unep.org.adaptationgapreport/files/documents/agr2016.pdf. Accessed 20 Jan 2017.

UNFCCC. 1992. *The United Nations Framework Convention on Climate Change*, in The United Nations Framework Convention on Climate Change *FCCC/INFORMAL/84*.

———. 1995. Report of The Conference of The Parties on Its First Session, Berlin, 28 March to 7 April 1995. FCCC/CP/1995/7/Add.1. Available at: http://unfccc.int/resource/docs/cop1/07a01.pdf. Accessed 24 Feb 2014.

———. 1996a. *Financial Mechanism, Memorandum of Understanding Between the Conference of the Parties and the Global Environment Facility*, in Subsidiary Body for Implementation (ed.), *FCCC/SBI/1996/L.4/Rev.1*, Geneva.

———. 1996b. *National Communications*, in Subsidiary Body for Scientific and Technological Advice and Subsidiary Body for Implementation and Ad Hoc Group on the Berlin Mandate *FCCC/AGBM/1996/MISC.1/ADD.1*, Geneva.

———. 1997. *Activities Implemented Jointly Under the Pilot Phase*, in Subsidiary Body for Scientific and Technological Advice *FCCC/SBSTA/1997/MISC.5*, Bonn.
———. 1998a. *Position Papers on Matters Before the Eighth Sessions of the Subsidiary Bodies*, in Subsidiary Body for Scientific and Technological Advice and Subsidiary Body for Implementation *FCCC/SB/1998/MISC.4*, Bonn.
———. 1998b. *Proposals on Financial Mechanism*, in Subsidiary Body for Implementation *FCCC/SBI/1998/MISC.4*, Bonn.
———. 2000. *Financial Mechanisms, Other Matters*, in Subsidiary Body for Implementation (ed.), *FCCC/SBI/2000/MISC.1*, Lyon.
———. 2002. *National Communications from Parties Not Included in Annex I to the Convention, Work of the Consultative Group of Experts*, in Subsidiary Body for Implementation *FCCC/SBI/2002/MISC.6*, New Delhi.
———. 2003. *Financial Matters Relating to Parties Not Included in Annex I to the Convention, Financial Mechanism: The Special Climate Change Fund*, in Subsidiary Body for Implementation *FCCC/SBI/2003/MISC.1*, Bonn.
———. 2004. *Exchange of Views on UNFCCC Activities Relevant to Other Intergovernmental Meetings, Submissions from Parties*, in Conference of the Parties to the United Nations Framework Convention on Climate Change (ed.), *FCCC/CP/2004/MISC.2*, Buenos Aires.
———. 2005. *Draft Decision Proposed by Jamaica on Behalf of the Group of 77 and China*, in Subsidiary Body for Implementation (ed.), *FCCC/SBI/2005/CRP.1*, Montreal.
———. 2006. *Views on Specific Policies, Programme Priorities and Eligibility Criteria and Possible Arrangements for the Management of the Adaptation Fund, in Subsidiary Body for Implementation* FCCC/SBI/2006/MISC.7, Bonn.
———. 2010. *Decision 1/CP.16*. http://unfccc.int/resource/docs/2010/cop16/eng/07a01.pdf#page=2. Accessed 15 Mar 2017.
———. 2014. *2014 Biennial Assessment and Overview of Climate Finance Flows Report*. UNFCCC Standing Committee on Finance. https://unfccc.int/files/cooperation_and_support/financial_mechanism/standing_committee/application/pdf/2014_biennial_assessment_and_overview_of_climate_finance_flows_report_web.pdf. Accessed 20 Aug 2016.
———. 2015. Historic Paris Agreement on Climate Change: 195 Nations Set Path to Keep Temperature Rise Well Below 2 Degrees Celsius [Online]. Available at: http://newsroom.unfccc.int/unfccc-newsroom/finale-cop21/. Accessed 12 Aug 2016.
UNICEF EAPRO Media Centre. 2011. *United Nations Secretary General Calls for Unity on Climate Change in Kiribati*. Available at: https://www.unicef.org/pacificislands/media_17174.html. Accessed 2 June 2013.

United Nations Department of Public Information. 2009. General Assembly, Expressing Deep Concern, Invites Major United Nations Organs to Intensify Efforts in Addressing Security implications of Climate Change. News Release GA/10830 3 June.

United Nations Security Council. 2007. *Security Council Holds First Ever Debate on Impact of Climate Change on Peace, Security Hearing Over 50 Speakers.* UN Security Council SC/9000.

———. 2011. *Statement by the President of the Council.* S/PRST/2011/15.

Van Schendel, W. 2002. Geographies of Knowing, Geographies of Ignorance: Jumping Scale in Southeast Asia. *Environment and Planning D: Society and Space* 20 (6): 647–668.

Wainwright, E. 2003. Responding to State Failure: The Case of Australia and the Solomon Islands. *Australian Journal of International Affairs* 57 (3): 485–498.

Watson, C., S. Patel, and L. Schalatek. 2016a. *Climate Finance Thematic Briefing: REDD+ Finance.* London/Washington, DC: ODI and Heinrich Böll Stiftung.

Watson, C., S. Patel, A. Durand, and L. Schalatek. 2016b. *Climate Finance Briefing: Small Island Developing States.* London/Washington, DC: ODI and Heinrich Böll Stiftung.

Webersik, Christian. 2010. *Climate Change and Security: A Gathering Storm of Global Challenges.* Santa Barbara: Praeger.

Weir, T., and Z. Virani. 2011. Three Linked Risks for Development in the Pacific Islands: Climate Change, Disasters and Conflict. *Climate and Development* 3 (3): 193–208.

Williams, M.C. 2003. Words, Images, Enemies: Securitization and International Politics. *International Studies Quarterly* 47 (4): 511–531.

World Bank. 2006. *An Investment Framework for Clean Energy and Development.* Washington, DC: World Bank.

———. 2009. *World Development Report 2010: Development and Climate Change.* Washington, DC: World Bank.

Wyeth, K. 2014. Escaping a Rising Tide: Sea Level Rise and Migration in Kiribati. *Asia and The Pacific Policy Studies* 1 (1): 171–185.

Yang, J. 2009. China in the South Pacific: Hegemon on the Horizon? *The Pacific Review* 22 (2): 139–158.

Index[1]

A
Aisi, Robert, 69
Alliance of Small Island States (AOSIS), 4, 6, 9, 25, 41, 43, 54–57, 102–105, 114
American Samoa, 5, 20, 30
Asian Development Bank (ADB), 16, 98
Australia, 4–6, 10, 14, 16, 18–22, 25, 27, 30, 31, 33, 44, 46, 56, 70, 71, 77, 81, 98, 113

B
Bani, Ernest, 55
Berlin Mandate, the (1995), 55

C
China, 52, 53, 57, 70, 102, 103
Clark, Helen, 22
Climate change adaptation, 45, 53, 79, 88, 90, 95
Climate change mitigation, 53, 88–90
Climate security, 2, 3, 5, 9, 10, 63–81, 109, 111, 112, 114
Commonwealth of the Northern Marianas, 20
Commonwealth, The, 20
Conference of the Parties (COP), through UNFCCC
  COP 15 Copenhagen 2009, 26, 56, 93, 96
  COP 16 Cancun 2010, 92
  COP 17 Durban 2011, 92
  COP 21 Paris 2015, 10, 56
Cook Islands, 4, 5, 17, 18, 20, 28–30, 32, 100
Copenhagen School, 66, 67, 73, 114
Coral Triangle Initiative, 19, 20
Council of Regional Organizations in the Pacific (CROP), 15, 21, 32, 105

[1] Note: Page number followed by 'n' refers to notes.

M. Williams, D. McDuie-Ra, *Combatting Climate Change in the Pacific*, https://doi.org/10.1007/978-3-319-69647-8

E

El Niño, 42
European Union (EU), 25, 27, 28, 33, 98
Extinction of Pacific Island states and societies, 43

F

Federated States of Micronesia, 5, 20, 27, 28, 30, 72
Fiji, 5, 6, 17–22, 28–30, 32, 35, 70, 100, 111, 113
Food and Agriculture Organization (FAO), 75
Food security, 10, 29, 44, 45, 68, 74–76
Forestry, 28, 112
  deforestation, 33
Framework for Pacific Regionalism, The, 23
Framework for Resilient Development in the Pacific (FRDP), 24, 25, 42
France, 4, 5, 14, 27, 30, 31, 98
French Polynesia, 4, 5, 20, 30, 35n2
Frontline of climate change, 2, 3, 9, 10, 40, 41, 53, 54, 57, 113, 114

G

G77, 54, 56, 57, 102, 103
Gender (and climate change), 45, 78
Global Environmental Facility (GEF), 32, 90–92, 98, 100, 103, 104
Green Climate Fund (GCF), 90–93, 98
Guam, 4, 5, 20, 27, 30

H

Health and climate change, 74
Hegemony, 9, 41, 51, 53, 113

I

Indonesia, 4, 19
Intergovernmental Panel on Climate Change (IPCC), 42, 43, 55, 71
International Organization for Migration, 20

K

Ki-Moon, Ban, 39, 88
Kiribati, 5, 9, 20, 22, 28–30, 32, 39, 40, 44, 47, 50, 54, 71, 102
Kyoto Protocol, the (1997), 43, 56, 90, 109, 110

L

Least Developed Countries Fund (LDCF), 91–93, 98
Levi, Noel, 19, 22
Lierop, Van, 43, 55

M

Malaysia, 19
Mara, Ratu Sir Kamisese, 17, 18
The Marshall Islands, 5, 44, 48
Media portrayals of climate change, 47
Melanesian Spearhead Group, 19, 113
Migration, 10, 44, 46–50, 52, 68, 69, 71, 72, 74, 75, 77, 78, 113
  climate refugees, 44
Migration, remittances, aid dependency and bureaucracy as characteristics of Pacific economies (MIRAB), 49
Montreal Protocol, the (1987), 55
Mori, Emmanuel, 72

N

Nauru, 5, 17, 20, 28, 30, 32, 72
The Netherlands, 4, 14

New Caledonia, 5, 20, 30, 32n2
New Zealand, 4–6, 10, 14, 16, 19–22, 27, 30, 31, 33, 35, 44, 56, 71, 77, 98, 113
Niue, 4, 5, 20, 24, 27, 28, 30, 32, 71, 75, 102
Niue Declaration on Climate Change (2008), 24
Northern Mariana Islands, 5, 30

O
Organization of economic Cooperation and Development (OECD), 76, 96, 98

P
Pacific Aviation Safety Office, 15
Pacific Community, 4, 11n1, 24, 26–27, 29, 34
  South Pacific Commission (SPC), 14, 26
The Pacific Islands Applied Geoscience Commission (SOPAC), 6, 25, 34
Pacific Islands Development Forum (PIDF), 111
Pacific Islands Development Programme (PIDP), 15
Pacific Islands Forum Fisheries Agency (FFA), 15, 16
Pacific Islands Forum (PIF), 4, 6, 9, 14, 15, 17, 18, 20–26, 33–35, 39, 71, 72, 105, 113
  Pacific Islands Forum Secretariat (PIFS), 18, 21, 22
Pacific Islands Framework for Action on Climate Change (PIFACC), 24, 25, 31
Pacific Plan, The, 19, 22, 23, 25
Pacific Power Association, 15
The 'Pacific Way', 8
Paeniu, Bikenibeu, 56
Palau, 4, 5, 20, 28, 30, 32, 72
Papua New Guinea, 4–6, 20–22, 28, 30, 32, 69, 72, 74, 75, 100
Paris Agreement 2015, 10, 11, 97, 110–113
  COP 21, 10
The Philippines, 4, 19
Pitcairn Islands, 5, 30
Policy narratives (of climate change), 3, 7, 35, 41–44, 46, 48, 57, 63
Poverty, 10, 44, 50, 52, 74, 76, 80, 95
Puna, Henry, 18

R
REDD, 28, 100
REDD+, 29, 95
  See also (REDD)

S
Samoa, 5, 6, 17, 20, 22, 28–32, 55, 100, 104
Sareer, Ahmed, 56
Secretariat of the Pacific Regional Environmental Program (SPREP), 6, 9, 15, 20, 24, 25, 29–34, 35n6, 102, 105
Securitization, 66, 67, 73, 114
Shared fate of Pacific states and territories viz. climate change, 35
Slade, Tuiloma Neroni, 18, 22
Solidarity in Pacific politics, 35
Solomon Islands, 5, 6, 17, 19, 20, 27–30, 32, 52, 70, 77, 100
South Korea, 52
South Pacific Tourism Organization (SPTO), 15
Suva Climate Declaration 2015, 111

T

Taiwan, 53, 70
Taylor, Dame Meg, 21, 22
Timor-Leste, 4, 19, 20
Tokelau, 4, 5, 20, 30, 44
Tonga, 5, 17, 20, 28–30, 32
Tourism, 15, 28, 33, 53, 76
Tukuitonga, Dr Colin, 27
Tuvalu, 5, 9, 16, 20, 22, 28, 30, 32, 40, 44, 47, 51, 54, 56, 57, 71, 77

U

The United Kingdom (UK), 4, 14, 31, 69, 98
United Nations Development Program (UNDP), 6, 32, 34, 50, 65, 90–92, 96, 100, 105
United Nations Economic and Social Commission for Asia and the Pacific (UNESCAP), 6, 15, 16, 29
United Nations Environmental Program (UNEP), 29, 43, 90, 94, 96
United Nations Framework Convention on Climate Change (UNFCCC), 8, 10, 25, 26, 32, 40, 54, 55, 88, 90, 93, 94, 96, 102–105, 110
United Nations General Assembly (UNGA), 17, 18, 55, 72, 102
United Nations Security Council (UNSC), 69, 72, 114
United States of America, 4
University of the South Pacific, 6, 15, 25, 32
Urbanization in the Pacific, 50, 51, 57, 75

V

Vanuatu, 5, 6, 20, 27–30, 32, 43, 55
Vulnerabilities (from climate change), 74

W

Wallis and Futuna, 5, 20, 30
Western and Central Pacific Fisheries Commission, 20
World Bank, The, 20, 90–92, 95, 96

CPSIA information can be obtained
at www.ICGtesting.com
Printed in the USA
LVHW081933090619
620633LV00012B/327/P